SpringerBriefs in Applied Sciences and Technology

Series Editor

Francis A. Kulacki
Department of Mechanical Engineering
Minneapolis, Minnesota, USA

Michael F. Modest • Daniel C. Haworth

Radiative Heat Transfer in Turbulent Combustion Systems

Theory and Applications

 Springer

Michael F. Modest
University of California
Merced, CA, USA

Daniel C. Haworth
Department of Mechanical and Nuclear
 Engineering
The Pennsylvania State University
State College, PA, USA

ISSN 2191-530X   ISSN 2191-5318   (electronic)
SpringerBriefs in Applied Sciences and Technology
ISBN 978-3-319-27289-4   ISBN 978-3-319-27291-7   (eBook)
DOI 10.1007/978-3-319-27291-7

Library of Congress Control Number: 2015957967

Springer Cham Heidelberg New York Dordrecht London

Springer International Publishing AG Switzerland is part of Springer Science+Business Media (www.
springer.com)

Preface

The first author has been involved in research dealing with thermal radiation in combustion systems for the past 30 years or so and first became interested in Turbulence–Radiation interactions in the mid-1990s. When the second author joined the Penn State faculty in 1999, they both formed a team - with Haworth providing expertise in CFD, turbulence, and combustion and Modest in the field of thermal radiation. This cooperation has prospered over the years, resulting in numerous joint research grants from NSF, NASA, DoE, NETL, and AFOSR and continues fruitfully 6 years after the first author moved on to the University of California, Merced. This monograph reviews our own work, and the work of others, on the effects of radiation in turbulent reactant flows and how to deal with these effects theoretically and numerically in simulations.

Merced, CA, USA   Michael F. Modest
State College, PA, USA   Daniel C. Haworth
September 2015

Contents

Nomenclature

a	Weight function for WSGG, SLW, and FSK methods
A	Area, m^2
$A_{l,f}$	Pre-exponential factor for forward reaction l
$b_{l,f}$	Temperature exponent for forward reaction l
C	Cross section, m^2
C^*	Nondimensional radiant power
c, c_0	Speed of light, (in vacuum), m/s
c_β	Molar concentration of species β, mol_β/m^3
c_P	Constant pressure specific heat, J/kg K
C_μ	$k - \varepsilon$ model constant
$C_\phi, C_{\phi\Delta}$	Mixing model constant for RANS/PDF (LES/PDF)
d	Spectral line spacing, cm^{-1}
Da	Damköhler number
E, E_b	Emissive power, blackbody emissive power, W/m^2
$E_{l,f}$	Activation energy for forward reaction l, J/mol
f	k-distribution function, cm
f_v	Volume fraction
f_ϕ	Composition joint PDF
F_s	Absorption line blackbody distribution functions (ALBDF)
F_{rad}	Ratio of emitted radiation to sum of emitted radiation and convective wall heat transfer
$g_i, \mathbf{g}$	Body force component (vector), m/s^2
g	Cumulative k-distribution
G	Incident radiation, W/m^2
G	Filter function for LES, m^{-3}
h	Enthalpy, J/kg
$\Delta h_{f,\alpha}^0$	Species α formation enthalpy, J/kg
Δh_c	Lower heating value, J/kg
I	Radiative intensity, W/m^2sr
I_n^m	Intensity coefficient function, W/m^2
$\mathbf{J}^\alpha$	Molecular diffusive flux vector for species α, kg_α/m^2 s
$\mathbf{J}^h$	Molecular diffusive flux vector for enthalpy, J/m^2 s

k	Turbulence kinetic energy, m^2/s^2
k	Absorption coefficient variable, cm^{-1}
k	Absorptive index
$k_{l,f}, k_{l,r}$	Forward (reverse) reaction rate for reaction l
l_t	Turbulence length scale, m
L	Geometric length, m
m	Complex index of refraction
$\dot{m}_F$	Fuel mass flow rate, kg/s
M_α	Species α chemical symbol
n	Refractive index
$\hat{\mathbf{n}}$	Unit normal vector
N	Number of gray gases, or order of P_N-approximation
N_S	Number of chemical species
p	Pressure, bar
P_n^m	Associated Legendre polynomial
Pr	Prandtl number
Pr_t	Apparent turbulence Prandtl number
$q, \mathbf{q}$	Heat flux (vector), W/m^2
$\dot{Q}_{rad}$	Total radiative heat loss), W
Q	Random variable
R_c	Local cone radius, m
R_u	Universal gas constant, $= 8.3145\,J/mol\,K$
R	Random number
$\mathscr{R}$	TRI correlations
Re	Reynolds number
s, s'	Distance along path, m
$\hat{\mathbf{s}}$	Unit direction vector
S	Spectral line strength, or source function
$S_{\alpha,chem}$	Local mass chemical production rate for species α, $kg_\alpha/m^3\,s$
S_{rad}	Local radiation source term in energy equation, W/m^3
S_{abs}	Absorption contribution to local radiation source term, W/m^3
S_{emi}	Emission contribution to local radiation source term, W/m^3
t	Time, s
T	Temperature, K
T_A	Activation temperature, K
u	Scaling function for absorption coefficient
$u_i, \mathbf{u}$	Velocity component (vector), m/s
V	Volume, m^3
w	Quadrature weight
W_α	Species α molecular weight, kg_α/mol_α
$W_i, \mathbf{W}$	Wiener process component (vector), $s^{1/2}$
$x, \mathbf{x}$	Location (vector), m
$Y, \mathbf{Y}$	Mass fraction (vector)
Y_n^m	Spherical harmonic, sr

Greek Symbols

β	Extinction coefficient, cm^{-1}
β	Spectral line overlap parameter
Δ	LES filter scale, m
ϵ	Gas emissivity, or wall emittance
ε	Turbulence dissipation rate, m^2/s^3
η	Wavenumber, cm^{-1}
$\Delta\eta$	Narrow band wavenumber range, cm^{-1}
γ	Spectral line half-width, cm^{-1}
$\Gamma_{T\phi}$	Apparent turbulence diffusivity, kg/m s^2
λ	Wavelength, μm
κ	Absorption coefficient, cm^{-1}
μ_T	Apparent turbulence viscosity, kg/m s^2
ν', ν''	Stoichiometric coefficients
$\boldsymbol{\phi}$	Composition variable vector
Φ	Scattering phase function
Φ	Viscous dissipation rate, W/m^3
ω	Scattering albedo
ω, ω_Δ	Turbulence mixing frequency for RANS (LES), s^{-1}
$\dot{\omega}_\alpha$	Local molar chemical production rate for species α, kmol$_\alpha$/m^3 s
Ω	Solid angle, sr
ρ	Density, kg/m^3
σ	Stefan–Boltzmann constant, $= 5.670 \times 10^{-8}$ W/m^2 K^4
σ_s	Scattering coefficient, cm^{-1}
σ_ϕ	Apparent turbulence Schmidt number or Prandtl number
θ, ψ	Polar and azimuthal angles, rad
$\boldsymbol{\psi}$	Sample space vector for random variables
τ	Gas transmissivity
τ	Optical depth
τ	Turbulence integral time scale, s
$\boldsymbol{\tau}$	Viscous stress tensor, Pa
χ_R	Radiant fraction

Subscripts

0	Reference state
abs	Absorption
b	Blackbody emission
chem	Chemical
emi	Emission
F, O, P	Fuel, oxidizer, product

g	Spectral, with cumulative k-distribution as spectral variable
i, j, k	Cartesian components
j	Narrow band
k	Spectral, with absorption coefficient as spectral variable
l	Reaction counter
max	Maximum
min	Minimum
mix	Mixture
rad	Radiative
w	Wall
α, β	Species counters
η, λ	Spectral

Overscores

$-$	Narrow-band average
$\sim$	Favre average, or spatial average narrow-band value
$\frown$	Density-weighted filtered value

Operators

| $\langle\ \rangle$ | Mean value |
| $\langle\ \rangle_\Delta$ | Spatially filtered value |

Acronyms

ADF	Absorption distribution function
ALBDF	Absorption line blackbody distribution function
CFD	Computational fluid dynamics
CPU	Central processing unit
EGR	Exhaust gas recirculation
FSK	Full-spectrum k-distribution
FSCK	Full-spectrum correlated k-distribution
FSSK	Full-spectrum scaled k-distribution
DNS	Direct numerical simulation
DOM	Discrete ordinates method
FVM	Finite volume method
HACA	Hydrogen abstraction C_2H_2 addition
IC	Internal combustion
IDF	Inverse diffusion flame
IEM	Interaction by exchange with the mean
LBL	Line by line
LES	Large eddy simulation
NDF	Normal diffusion flame
OTFA	Optically thin fluctuation assumption
PAH	Polycyclic aromatic hydrocarbon
PDE	Partial differential equation
PDF	Probability density function
PM	Particulate matter
PMC	Photon Monte Carlo
PSDF	Particle size distribution function
RANS	Reynolds-averaged Navier-Stokes
RTE	Radiative transfer equation
SHM	Spherical harmonics method
SLW	Spectral-line weighted sum of gray gases
SNB	Statistical narrow band
SNBCK	Statistical narrow band correlated-k

TCI	Turbulence–chemistry interaction
TRI	Turbulence–radiation interaction
WSGG	Weighted sum of gray gases
WSR	Well-stirred reactor

Chapter 1
Introduction

1.1 Motivation

Thermal radiation, chemical kinetics, and turbulence individually are among the most challenging fundamental and practical problems of computational science and engineering. In turbulent reacting flows, these phenomena are coupled in interesting and highly nonlinear ways, leading to entirely new classes of interactions. Coupling between turbulence and chemistry [turbulence–chemistry interaction: (TCI)] has received considerable attention in the turbulent combustion literature for many years, and multiple review papers, monographs and books have been published on the subject [1–11]. Turbulence influences transport processes in flames through stretch and curvature, for example, and heat release in turn influences the turbulent flow field through changes in the temperature-dependent properties. TCI influences the local rate of heat release in a turbulent flame, and can result in local or even global extinction. Emissions of key pollutants have been shown to change by orders of magnitude in some cases with consideration of TCI.

Little attention had been paid until recently to accurate modeling of radiative heat transfer in combustion systems, in spite of the fact that radiation is often the dominant mode of heat transfer (being ignored or treated with simplistic "optically thin" and "gray" models) [12, 13], primarily perhaps because including accurate radiation calculations would add another layer of difficulty to an already daunting problem. However, during the past few years it has been widely recognized that neglecting radiation in atmospheric pressure combustion systems may lead to *overprediction* of temperature of up to 200 °C, while using the usually employed optically thin or gray radiation models leads to *underprediction* of up to 100 °C and more [14, 15]. Radiation and spectral radiation properties significantly alter the propagation speed and extinction characteristics of laminar premixed flames [16]. Measurements in laboratory-scale nonluminous, nonpremixed turbulent methane jet flames have shown that emission-only (optically thin) radiation models can

© The Author(s) 2016
M.F. Modest, D.C. Haworth, *Radiative Heat Transfer in Turbulent Combustion Systems*, SpringerBriefs in Applied Sciences and Technology,
DOI 10.1007/978-3-319-27291-7_1

overpredict the radiant heat loss by more than a factor of two, with commensurately large errors in NO predictions [17]; product-gas (CO_2) spectral radiation properties are responsible for this behavior.

Furthermore, while the importance of turbulence–radiation interaction (hereafter TRI) has long been recognized [18–22], until recently TRI has received relatively little attention by turbulent combustion researchers. TRI can lead to significant changes in local and global flame properties through highly nonlinear interactions among turbulent fluctuations in temperature and composition and, hence, radiative intensity. Changes in mean temperature, heat transfer rates, and pollutant emissions resulting from TRI can be comparable to or greater than those resulting from TCI in practical combustion systems [20–22]. For example, experimental measurements in laboratory-scale flames [23–30] have shown that radiative emission can be 50–300 % higher than that expected based on mean values of temperature and absorption coefficient, depending on the fuel. Modeling studies of nonluminous flames [14, 31–33] have reported increases of 30–50 % in radiative emission and reductions in peak mean temperature exceeding 100 K with consideration of TRI compared to simulations that consider radiation but no TRI. Coelho [34] has recently provided an exhaustive review of TRI in turbulent combustion, with emphasis on numerical simulations.

In higher-pressure, larger-scale, and/or luminous (sooting) turbulent flames the effects of radiation, spectral radiation properties, and TRI can be even more pronounced. For example, in CFD simulations of in-cylinder processes in piston engines, the influences of radiation processes on heat transfer and heat loss estimations have been largely ignored, in spite of the potentially strong emission and reabsorption from hot combustion gases and particles (soot). Recent evidence (see Chap. 7) shows that radiation can contribute significantly to redistributing energy in the combustion chamber and to wall heat losses in engines, thus leading to considerably different computed temperature distributions and emissions levels when radiation is considered.

1.2 Reynolds Averaging, Spatial Filtering, and Probability Density Function Methods

A wide dynamic range of length and time scales characterizes chemically reacting turbulent flows at the high Reynolds (Re) and Damköhler (Da) numbers (the latter being the ratio of a characteristic flow or turbulence time scale to a characteristic chemical time scale) that are typical of engineering turbulent combustion systems. The largest relevant length scales are fixed by the size of the device or system of interest, while the smallest may correspond to the smallest hydrodynamic scales (the Kolmogorov turbulence microscale), to a chemical or flame scale (e.g., the local flame thickness), or to scales that characterize other physical processes, such as droplet sizes in liquid fuel sprays. For CFD simulations, the range of scales must be reduced to a computationally tractable level, and this can be accomplished using probabilistic approaches or filtering. In Reynolds-averaged Navier–Stokes (RANS), *all* fluctuations about the probabilistic mean (estimated as an ensemble average over statistically equivalent flow realizations, a spatial average over directions of statistical homogeneity, and/or a time average in statistically stationary flows) are modeled. By contrast, in large-eddy simulation (LES) the governing equations are spatially filtered [35–39]; the large-scale dynamics (large turbulent eddies whose transient behavior can be resolved numerically) are captured explicitly, while the effects of unresolved (subfilter) scales are modeled. LES is expected to be more accurate and general compared to RANS, since at least some of the fluctuations (the largest, most energetic ones) are captured explicitly and only the (presumably) more universal small-scale dynamics require modeling. LES is also expected to capture phenomena that are difficult to accommodate in Reynolds-averaged approaches, such as large-scale unsteadiness (e.g., combustion instabilities in gas-turbine combustors, cycle-to-cycle variations in reciprocating-piston IC engines). LES is inherently three-dimensional and time-dependent, while RANS can take advantage of statistical homogeneity and/or stationarity to reduce the number of independent variables.

There have been rapid advances in LES for chemically reacting turbulent flows over the past 10–15 years. The review by Pitsch [40] includes examples of applications, discussion of issues specific to reacting flows, and important differences between RANS and LES. There is a wide and rapidly growing body of evidence that demonstrates quantitative advantages of LES in modeling studies of laboratory flames [41–43] and in applications to gas-turbine combustors [44–47], piston engines [48–50], and other combustion systems. Several LES studies focusing on radiative heat transfer in turbulent combustion have appeared recently, and these will be discussed in later chapters. However, LES remains computationally intensive compared to RANS, and RANS is still of interest for practical applications. Both RANS and LES will be considered here.

Probability density function (PDF) methods provide an effective framework for describing and modeling the influences of fluctuations with respect to local mean values in RANS, or of subfilter-scale fluctuations with respect to local spatially filtered values in LES. In this approach, one considers the probabilities that the fluctuations in quantities of interest have particular values, which are quantified by introducing a PDF, and a modeled transport equation is solved for the PDF of interest. For example, to compute the local instantaneous rate of radiative emission $S_{\mathrm{emi}}(\mathbf{x}, t)$ in a mixture of molecular gases, one needs to know the local partial pressures of the participating species (or equivalently, the mixture pressure and the mass or mole fractions of the participating species) and the local temperature (see Chap. 3): $S_{\mathrm{emi}}(\mathbf{x}, t) = S_{\mathrm{emi}}\big(p(\mathbf{x}, t), \mathbf{Y}(\mathbf{x}, t), T(\mathbf{x}, t)\big)$, where p is the mixture pressure, $\mathbf{Y}$ is the vector of mass fractions of participating species, and T is the temperature. In RANS, one has access only to mean quantities (denoted using angled brackets $\langle\ \rangle$), and the local mean rate of radiative emission $(\langle S_{\mathrm{emi}}(\mathbf{x}, t)\rangle)$ is needed. Because the dependence of $S_{\mathrm{emi}}(\mathbf{x}, t)$ on temperature and composition is nonlinear, $\langle S_{\mathrm{emi}}(\mathbf{x},t)\rangle \neq S_{\mathrm{emi}}(\langle p(\mathbf{x},t)\rangle, \langle \mathbf{Y}(\mathbf{x},t)\rangle, \langle T(\mathbf{x},t)\rangle)$. On the other hand, if one had access to the joint PDF of p, $\mathbf{Y}$ and T, denoted $f_\phi(\boldsymbol{\psi};\mathbf{x},t)$ where ϕ is the vector of *composition variables* needed to compute the local instantaneous value of $S_{\mathrm{emi}}(\mathbf{x},t)$ (here p, $\mathbf{Y}$ and T) and $\boldsymbol{\psi}$ is the corresponding sample space vector, then $\langle S_{\mathrm{emi}}(\mathbf{x},t)\rangle = \int S_{\mathrm{emi}}(\boldsymbol{\psi})f_\phi(\boldsymbol{\psi};\mathbf{x},t)d\boldsymbol{\psi}$, where integration is over the entire sample space. In a PDF method, then, the emphasis shifts from attempting to express the mean radiative emission rate in terms of other mean quantities to computing the joint PDF of the quantities that uniquely determine the local instantaneous radiative emission rate. Similar arguments apply to the mean chemical source terms, and to spatially filtered quantities in LES.

PDF methods for dealing with mean (or filtered) chemical source terms have been developed and applied extensively in turbulent combustion research and applications over several decades, and more recently PDF methods have been adopted for modeling radiative heat transfer in turbulent combustion. The foundations of PDF methods for turbulent reacting flows can be found in [51], progress through approximately 2009 is reviewed in [52], and recent developments and trends are discussed in [53]. PDF methods are adopted here as the framework for elucidating and modeling the influences of unresolved turbulent fluctuations on chemical and radiative processes in turbulent combustion, and the essential elements of PDF methods in RANS and LES for present purposes are outlined in Chaps. 2 and 3 below.

1.3 Chemical Reactions, Reaction Mechanisms, and Turbulence–Chemistry Interaction

As was the case for the radiative emission S_{emi} in the example discussed above, the chemical source term $S_{\alpha,\mathrm{chem}}$ in the unaveraged/unfiltered conservation equation for each chemical species α in a multicomponent reacting mixture containing N_S different species is (at least, in principle) a known function of the local mixture pressure p, the local mixture composition (which can be expressed in terms of the vector of the N_S species mass fractions $\mathbf{Y}$), and the local temperature T:

$$S_{\alpha,\mathrm{chem}}(\mathbf{x},t) = S_{\alpha,\mathrm{chem}}(p(\mathbf{x},t), \mathbf{Y}(\mathbf{x},t), T(\mathbf{x},t)) . \tag{1.1}$$

A reaction mechanism provides the specific functional form. A detailed reaction mechanism (a set of reactions that describes the outcomes of all molecular collisions of interest) even for a single hydrocarbon fuel molecule reacting with oxygen can involve hundreds to thousands of individual reactions among hundreds of individual chemical species. For example, a recent reaction mechanism for n-heptane (sometimes used as a single-component surrogate for diesel fuel) considers 654 species and 5935 reactions [54]. Even larger numbers of species and reactions are required to accurately describe the combustion of real multicomponent fuels. In practice, skeletal or reduced mechanisms involving up to ~100 species, say, are usually used for CFD of turbulent combustion. Such simplified mechanisms are restricted to relatively narrow ranges of thermochemical conditions (pressures, temperatures, and equivalence ratios).

Individual reaction rates are usually expressed using a modified Arrhenius form, as discussed in Sect. 2.1 below. The reaction rates then exhibit up to cubic nonlinearities in species compositions (since up to three-body reactions are of interest in combustion) and an exponential nonlinearity in the temperature. A consequence of these nonlinearities is that in RANS (LES), the mean (filtered) chemical source term cannot be expressed simply in terms of the mean (filtered) pressure, mean (filtered) species mass fractions, and the mean (filtered) temperature. In RANS:

$$\langle S_{\alpha,\text{chem}}(\mathbf{x}, t) \rangle \neq S_{\alpha,\text{chem}}\big(\langle p(\mathbf{x}, t) \rangle, \langle \mathbf{Y}(\mathbf{x}, t) \rangle, \langle T(\mathbf{x}, t) \rangle \big) , \qquad (1.2)$$

and a similar inequality holds for the spatially filtered chemical source terms in LES. A major difficulty in modeling chemically reacting turbulent flows in RANS (LES) is that contributions of unresolved turbulent fluctuations to the mean (filtered) chemical source terms generally are not negligible; this is a manifestation of TCI. Neglect of these interactions may result in overestimation of local heat-release rates and peak mean combustion temperatures (by as much as several 100 °C), inaccurate emissions predictions (by >100 %), and failure to capture local and global ignition/extinction phenomena. TCI and the development of closure models for mean (filtered) chemical source terms have been the subjects of extensive research, and have been covered in review articles [55] and books [4, 6, 56, 57].

An initial example illustrating the importance of TCI in a practical turbulent combustion system is provided in Fig. 1.1. There CFD (unsteady RANS) computed soot emissions are compared with experimental particulate matter measurements for a heavy-duty direct-injection diesel engine. Results from two different TCI modeling treatments are shown, with all other aspects of the CFD models being the same. In one case, TCI have been ignored altogether by computing the local mean chemical source terms using the local mean pressure, composition, and temperature: a locally well-stirred reactor (WSR) model at the CFD cell level. In the other case, a transported composition PDF method (Chap. 2) has been used to account for the effects of unresolved turbulent fluctuations with respect to the local mean values. The WSR model yields a burn rate that is too fast, and predicts that essentially all of the in-cylinder soot that is generated during the combustion event is oxidized before the exhaust-valve opens. The computed soot levels from the PDF model are up to *five orders of magnitude higher* than those from the WSR model, and are in much closer agreement with experiment; the reasons for these very large differences are explored in detail in [58]. Other examples will be discussed in subsequent chapters, after the theoretical underpinnings have been established.

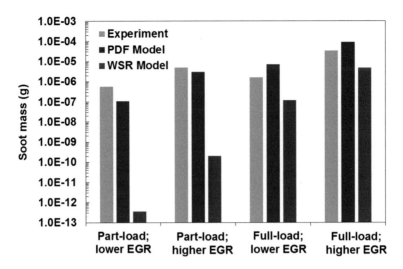

Fig. 1.1 Computed and measure particulate (soot) emissions for a heavy-duty diesel engine at four different operating conditions [58]

1.4   Thermal Radiation, Spectral Radiation Properties, and Turbulence–Radiation Interaction

Consideration of thermal radiation introduces three new difficulties, in addition to the already formidable issues of turbulence, chemical kinetics, and TCI. First, the absorption coefficient κ_η exhibits strong and erratic variations with wavenumber η for the participating species that are of interest in combustion (principally H_2O, CO_2, and CO). Second, in cases where reabsorption is important, one must solve the radiative transfer equation (RTE) to obtain the local radiative intensity. The RTE is an integro-differential equation in five independent variables whose structure is quite different from that of the other governing equations for chemically reacting turbulent flows: special approaches are required to solve the RTE. And third, turbulence–radiation interactions (absorption TRI, in particular) are difficult to model.

An example for the absorption coefficient of combustion products is given in Fig. 1.2, showing a small part of the spectrum for the important 4.3 μm band of carbon dioxide, consisting of many partially overlapping spectral lines.

The radiation source term in the instantaneous energy equation, S_{rad}, is the difference between local absorption and local emission,

$$S_{\text{rad}} = -\nabla \cdot \mathbf{q}_{\text{rad}} = -\int_0^\infty \kappa_\eta \left(4\pi I_{b\eta} - \int_{4\pi} I_\eta d\Omega \right) d\eta$$

$$= -4\kappa_P \sigma T^4 + \int_0^\infty \int_{4\pi} \kappa_\eta I_\eta d\Omega d\eta . \qquad (1.3)$$

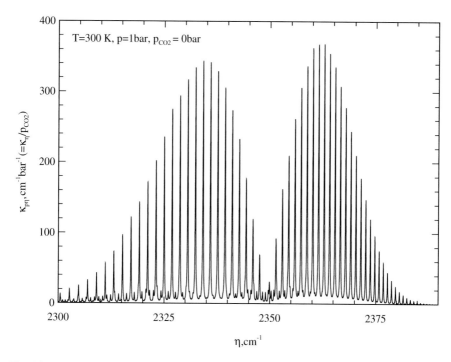

Fig. 1.2 Pressure-based spectral absorption coefficient for small amounts of CO_2 in nitrogen; 4.3 μm band at $p = 1.0\,\text{bar}$, $T = 300\,\text{K}$[59]

Here $\mathbf{q}_{\text{rad}}$ is the radiative heat flux, Ω is solid angle, κ_η is the spectral absorption coefficient ($\kappa_\eta = \kappa_\eta(\mathbf{Y}, T, p)$), $I_{b\eta}$ is the Planck function or blackbody intensity (a function of temperature and wavenumber), κ_P is the Planck-mean absorption coefficient, σ is the Stefan–Boltzmann constant, and I_η is the spectral radiative intensity. The intensity is obtained from the RTE [59]:

$$\frac{dI_\eta}{ds} = \hat{s} \cdot \nabla I_\eta = \kappa_\eta I_{b\eta} - (\kappa_\eta + \sigma_{s\eta})I_\eta + \frac{\sigma_{s\eta}}{4\pi} \int_{4\pi} I_\eta(\hat{s}_i)\Phi_\eta(\hat{s}_i, \hat{s})d\Omega_i, \qquad (1.4)$$

which states that spectral intensity along a path s is augmented by emission, diminished by absorption and outscattering (scattering away from the direction of propagation), and augmented by inscattering (scattering from other directions into the direction of propagation). Here $\hat{s}$ and $\hat{s}_i$ denote unit direction vectors, $\sigma_{s\eta}$ is the spectral scattering coefficient, and $\Phi_\eta(\hat{s}_i, \hat{s})$ denotes the scattering phase function. The local value of I_η depends on nonlocal quantities, on direction, and on wavenumber.

Much in the same way as convection is enhanced by turbulence through nonlinear interactions of velocity and temperature fluctuations, radiation is also enhanced by turbulence, by nonlinear interactions of temperature and radiative property fluctuations, which govern the strength of radiative intensity. In a Reynolds-averaged context, TRIs are brought into evidence by taking the mean of Eq. (1.3):

$$- S_{\text{rad}} = \langle \nabla \cdot \mathbf{q}_{\text{rad}} \rangle = \int_0^\infty \left(4\pi \langle \kappa_\eta I_{b\eta} \rangle - \int_{4\pi} \langle \kappa_\eta I_\eta \rangle d\Omega \right) d\eta. \tag{1.5}$$

Because of their nonlinear dependence on composition variables, these terms cannot be determined based on mean values. Thus, two *turbulence moments* or *correlations* are required: the correlations between absorption coefficient and Planck function, $\langle \kappa_\eta I_{b\eta} \rangle$, and between absorption coefficient and radiative intensity, $\langle \kappa_\eta I_\eta \rangle$. The former correlation is termed

$$\text{Emission TRI}: \qquad \langle \kappa_\eta I_{b\eta} \rangle \neq \kappa_\eta(\langle \boldsymbol{\phi} \rangle) I_{b\eta}(\langle T \rangle), \tag{1.6}$$

while the latter is known as

$$\text{Absorption TRI}: \qquad \langle \kappa_\eta I_\eta \rangle \neq \kappa_\eta(\langle \boldsymbol{\phi} \rangle) I_\eta(\langle \boldsymbol{\phi} \rangle). \tag{1.7}$$

Absorption TRI is particularly difficult to evaluate because the fluctuations of the local intensity may be influenced by property fluctuations from everywhere in the medium.

In LES a low-pass spatial filter is applied to the instantaneous governing equations [35, 37]. The instantaneous value of any physical quantity $Q = Q(\mathbf{x}, t)$ can be decomposed into a filtered component (denoted by $\langle \, \rangle_\Delta$, where Δ is the filter scale) and a fluctuation about the filtered component (the residual field, denoted by a prime): $Q(\mathbf{x}, t) = \langle Q(\mathbf{x}, t) \rangle_\Delta + Q'(\mathbf{x}, t)$, where $\langle Q \rangle_\Delta = \langle Q(\mathbf{x}, t) \rangle_\Delta \equiv \int Q(\mathbf{y}, t) G(|\mathbf{x} - \mathbf{y}|) d\mathbf{y}$, and integration is over the entire flow domain. The low-pass spatial filter function, $G(|\mathbf{x} - \mathbf{y}|)$ has a characteristic filter width Δ, and other properties are discussed in Sect. 2.2.2. In the limit as the filter scale Δ approaches l_T, where l_T is a length scale representative of the largest turbulent eddies, essentially all fluctuations are at subfilter scales and LES approaches (operationally) unsteady RANS.

An example showing a jet flame is provided in Fig. 1.3. There detailed chemistry, a transported composition PDF method, a photon Monte Carlo method for the solution of the radiative transfer equation, and spectral radiation properties of participating gas-phase mixtures (H_2O and CO_2) are combined to investigate TRI in piloted nonpremixed methane–air jet flames. The flame's maximum local mean temperature without radiation is approximately 1950 K; with consideration of nongray radiation but no TRI the temperatures drop about 150 K throughout the combustion zone; and accounting for TRI drops the temperature levels by another 50 K (the TRI changes are somewhat subtle for this relatively small flame, indicated by slightly smaller maximum temperature zones and slight shortening of the flame). Primitive radiation models, such as "optically thin radiation" (emission only) or treating the combustion gases as gray reduces the temperature by 300 K from the no-radiation case.

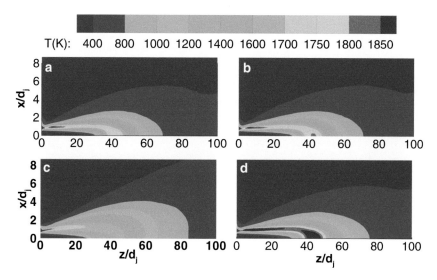

Fig. 1.3 Computed mean temperature contours for a scaled-up version of Sandia Flame D [60]. (**a**) With TRI. (**b**) Without TRI. (**c**) Gray radiation. (**d**) Without radiation

1.5 Organization of Subsequent Chapters

The remainder of the monograph is organized as follows. Chapters 2 and 3 provide further background about chemically reacting turbulent flows and radiative heat transfer, respectively. This includes further discussion of TCI and TRI, and an overview of the PDF-based modeling framework. Radiation effects can be important even in the absence of turbulence, and examples illustrating the importance of radiative transfer in laminar flames are provided in Chap. 4. In Chaps. 5–7, respectively, modeling and experimental studies of radiation and TRIs in canonical systems, atmospheric-pressure laboratory flames, and high-pressure turbulent combustion systems are presented and discussed. A summary, conclusions, and a perspective on future prospects are provided in the final chapter.

References

1. P.A. Libby, F.A. Williams, *Turbulent Reacting Flows* (Academic Press, San Diego, 1994)
2. R.W. Bilger, Turbulent diffusion flames. Ann. Rev. Fluid Mech. **21**, 101–135 (1989)
3. R.W. Bilger, Future progress in turbulent combustion research. Prog. Energy Combust. Sci. **26**, 367–380 (2000)
4. N. Peters, *Turbulent Combustion* (Cambridge University Press, Cambridge, 2000)
5. D. Veynante, L. Vervisch, Turbulent combustion modeling. Prog. Energy Combust. Sci. **28**, 193–266 (2002)

6. R.O. Fox, *Computational Models for Turbulent Reacting Flows* (Cambridge University Press, Cambridge, 2003)
7. C.K. Westbrook, Y. Mizobuchi, T.J. Poinsot, P.J. Smith, J. Warnatz, Computational combustion. Proc. Combust. Inst. **30**, 125–157 (2005)
8. R.S. Cant, E. Mastorakos, *An Introduction to Turbulent Reacting Flows* (Imperial College Press, London, 2008)
9. T. Poinsot, D. Veynante, *Theoretical and Numerical Combustion*, 3rd edn. (R.T. Edwards, Inc., Philadelphia, 2011)
10. T. Echekki, E. Mastorakos (eds.), *Turbulent Combustion Modeling – Advances, New Trends and Perspectives* (Springer, New York, 2011)
11. L. Vervisch, D. Veynante, J.P.A.J. Van Beeck (eds.), *Turbulent Combustion*. von Karman Institute for Fluid Dynamics Lecture Series 2013–06, Rhode-Saint-Genèse (2013)
12. W. Jones, M.C. Paul, Combination of DOM with LES in a gas turbine combustor. Int. J. Eng. Sci. **43**(5–6), 379–397 (2005)
13. P. Schmitt, T. Poinsot, B. Schuermans, K.P. Geigle, Large-eddy simulation and experimental study of heat transfer, nitric oxide emissions and combustion instability in a swirled turbulent high-pressure burner. J. Fluid Mech. **570**, 17–46 (2007)
14. S. Mazumder, M.F. Modest, A PDF approach to modeling turbulence–radiation interactions in nonluminous flames. Int. J. Heat Mass Transfer **42**, 971–991 (1999)
15. G. Li, M.F. Modest, Importance of turbulence–radiation interactions in turbulent diffusion jet flames. J. Heat Transfer **125**, 831–838 (2003)
16. Y. Ju, G. Masuya, P.D. Ronney, Effects of radiative emission and absorption on the propagation and extinction of premixed gas flames. Proc. Combust. Inst. **27**, 2619–2626 (1998)
17. J.H. Frank, R.S. Barlow, C. Lundquist, Radiation and nitric oxide formation in turbulent non-premixed jet flames. Proc. Combust. Inst. **28**, 447–454 (2000)
18. A.A. Townsend, The effects of radiative transfer on turbulent flow of a stratified fluid. J. Fluid Mech. **4**, 361–375 (1958)
19. G.M. Shved, R.A. Akmayev, Influence of radiative heat transfer on turbulence in planetary atmospheres. Atmos. Oceanic Phys. **34**, 1286–1401 (1977)
20. T.-H. Song, R. Viskanta, Interaction of radiation with turbulence: application to a combustion system. J. Thermophys. Heat Transfer **1**(1), 56–62 (1987)
21. A. Soufiani, P. Mignon, J. Taine, Radiation–turbulence interaction in channel flows of infrared active gases, in *Proceedings of the Ninth International Heat Transfer Conference*, vol. 6 (Hemisphere, Washington, D.C., 1990), pp. 403–408
22. R.J. Hall, A. Vranos, Efficient calculations of gas radiation from turbulent flames. Int. J. Heat Mass Transfer **37**, 2745 (1994)
23. J.P. Gore, G.M. Faeth, Structure and spectral radiation properties of turbulent ethylene/air diffusion flames, in *Proceedings of the Twenty-First Symposium (International) on Combustion*, pp. 1521–1531 (1986)
24. J.P. Gore, S.M. Jeng, G.M. Faeth, Spectral and total radiation properties of turbulent carbon monoxide/air diffusion flames. J. Am. Inst. Aeronaut. Astronaut. **25**(2), 339–345 (1987)
25. J.P. Gore, S.M. Jeng, G.M. Faeth, Spectral and total radiation properties of turbulent hydrogen/air diffusion flames. J. Heat Transfer **109**, 165–171 (1987)
26. J.P. Gore, G.M. Faeth, Structure and spectral radiation properties of luminous acetylene/air diffusion flames. J. Heat Transfer **110**, 173–181 (1988)
27. M.E. Kounalakis, J.P. Gore, G.M. Faeth, Turbulence/radiation interactions in nonpremixed hydrogen/air flames, in *Twenty-Second Symposium (International) on Combustion* (The Combustion Institute, Pittsburg, 1988), pp. 1281–1290
28. M.E. Kounalakis, J.P. Gore, G.M. Faeth, Mean and fluctuating radiation properties of non-premixed turbulent carbon monoxide/air flames. J. Heat Transfer **111**, 1021–1030 (1989)
29. M.E. Kounalakis, Y.R. Sivathanu, G.M. Faeth, Infrared radiation statistics of nonluminous turbulent diffusion flames. J. Heat Transfer **113**, 437–445 (1991)

30. Y.R. Sivathanu, M.E. Kounalakis, G.M. Faeth, Soot and continuous radiation statistics of luminous turbulent diffusion flames, in *Twenty-Third Symposium (International) on Combustion* (The Combustion Institute, Pittsburg, 1990), pp. 1543–1550

31. G. Li, M.F. Modest, Application of composition PDF methods in the investigation of turbulence–radiation interactions. J. Quant. Spectrosc. Radiat. Transfer **73**, 461–472 (2002)

32. G. Li, M.F. Modest, Importance of turbulence–radiation interactions in turbulent diffusion jet flames. J. Heat Transfer **125**, 831–838 (2003)

33. G. Li, M.F. Modest, Numerical simulation of turbulence–radiation interactions in turbulent reacting flows, in *Modelling and Simulation of Turbulent Heat Transfer*, ed. by B. Sundeń, M. Faghri, Chap. 3 (WIT Press, Southampton, 2005), pp. 77–112

34. P.J. Coelho, Numerical simulation of the interaction between turbulence and radiation in reactive flows. Prog. Energy Combust. Sci. **33**(4), 311–383 (2007)

35. A. Leonard, Energy cascade in large eddy simulation of turbulent flow. Adv. Geophys. **18A**, 237–248 (1974)

36. C. Meneveau, J. Katz, Scale-invariance and turbulence models for large-eddy simulation. Ann. Rev. Fluid Mech. **32**, 1–32 (2000)

37. S.B. Pope, *Turbulent Flows* (Cambridge University Press, Cambridge, 2000)

38. P. Sagaut, *Large Eddy Simulation for Incompressible Flows* (Springer, Berlin, 2001)

39. S.B. Pope, Ten questions concerning the large-eddy simulation of turbulent flows. New J. Phys. **6**(35) (2004). http://www.njp.org/

40. H. Pitsch, Large-eddy simulation of turbulent combustion. Ann. Rev. Fluid Mech. **38**, 453–482 (2006)

41. H. Pitsch, Improved pollutant predictions in large-eddy simulations of turbulent non-premixed combustion by considering scalar dissipation rate fluctuations. Proc. Combust. Inst. **29**, 1971–1978 (2002)

42. V. Raman, H. Pitsch, R.O. Fox, Hybrid large eddy simulation/Lagrangian filtered density function approach for simulating turbulent combustion. Combust. Flame **143**, 56–58 (2005)

43. M.R.H. Sheikhi, T.G. Drozda, P. Givi, F.A. Jaberi, S.B. Pope, Large eddy simulation of a turbulent nonpremixed piloted methane jet flame (Sandia Flame D). Proc. Combust. Inst. **30**, 549–556 (2005)

44. L. Selle, G. Lartigue, T. Poinsot, R. Koch, K.-U. Schildmacher, W. Krebs, B. Prade, P. Kaufmann, D. Veynante, Compressible large-eddy simulation of turbulent combustion in complex geometry on unstructured meshes. Combust. Flame **137**, 489–505 (2004)

45. P. Moin, S.V. Apte, Large-eddy simulation of realistic gas turbine combustors. AIAA Paper no. AIAA-2004-330 (2004)

46. S. James, J. Zhu, M.S. Anand, Large-eddy simulations of gas turbine combustors. AIAA Paper no. 2005-0552 (2005)

47. P. Flohr, CFD modeling for gas turbine combustors, in *Turbulent Combustion*, ed. by L. Vervisch, D. Veynante, J.P.A.J. Van Beeck. von Karman Institute for Fluid Dynamics Lecture Series 2005–02, Rhode-Saint-Genèse (2005)

48. I. Celik, I. Yavuz, A. Smirnov, Large-eddy simulations of in-cylinder turbulence for IC engines: a review. Int. J. Engine Res. **2**, 119–148 (2001)

49. D.C. Haworth, A review of turbulent combustion modeling for multidimensional in-cylinder CFD. SAE Trans. J. Engines 899–928 (2005)

50. S. Richard, O. Colin, O. Vermorel, A. Benkenida, A. Angelberger, D. Veynante, Towards large-eddy simulation of combustion in spark-ignition engines. Proc. Combust. Inst. **31**, 3059–3066 (2007)

51. S.B. Pope, PDF methods for turbulent reactive flows. Prog. Energy Combust. Sci. **11**, 119–192 (1985)

52. D.C. Haworth, Progress in probability density function methods for turbulent reacting flows. Prog. Energy Combust. Sci. **36**, 168–259 (2010)

53. D.C. Haworth, S.B. Pope, Transported probability density function methods for Reynolds-averaged and large-eddy simulations, in *Turbulent Combustion Modeling - Advances, New Trends and Perspectives*, ed. by T. Echekki, E. Mastorakos (Springer, Berlin, 2011), pp. 119–142
54. M. Mehl, W.J. Pitz, C.K. Westbrook, H.J. Curran, Kinetic modeling of gasoline surrogate components and mixtures under engine conditions. Proc. Combust. Inst. **33**, 193–200 (2011)
55. R.W. Bilger, S.B. Pope, K.N.C. Bray, J.F. Driscoll, Paradigms in turbulent combustion research. Proc. Combust. Inst. **30**, 21–42 (2005)
56. J. Warnatz, U. Maas, R.W. Dibble, *Combustion* (Springer, Berlin, 2001)
57. T. Poinsot, D. Veynante, *Theoretical and Numerical Combustion*, 2nd edn. (R.T. Edwards, Inc., Toulouse, 2005)
58. V.R. Mohan, D.C. Haworth, Turbulence-chemistry interactions in a heavy-duty compression-ignition engine. Proc. Combust. Inst. **35**, 3053–3060 (2015)
59. M.F. Modest, *Radiative Heat Transfer*, 3rd edn. (Academic Press, New York, 2013)
60. A. Wang, M.F. Modest, D.C. Haworth, L. Wang, Monte Carlo simulation of radiative heat transfer and turbulence interactions in methane/air jet flames, in *Radiative Transfer 2007 — The Fifth International Symposium on Radiative Transfer*, ed. by M.P. Mengüç, N. Selçuk (Begell House, Bodrum, 2007)

Chapter 2
Chemically Reacting Turbulent Flows

The equations governing a turbulent reacting system are presented, starting in
Sect. 2.1 with the unaveraged/unfiltered partial differential equations (PDEs) and
ancillary equations for a gas-phase multicomponent reacting system. These equations describe both laminar flames and turbulent flames where all continuum spatial
and temporal scales are fully resolved [direct numerical simulation—(DNS)]. In
simulations of practical turbulent combustion systems, it is not feasible to resolve
all relevant scales explicitly, and one of two approaches is usually adopted to
reduce the dynamic range of scales and to account for influences of turbulent
fluctuations at unresolved scales on the resolved scales (Sect. 2.2): Reynolds
averaging [Reynolds-averaged Navier–Stokes—(RANS)], in which the influences
of all turbulent fluctuations with respect to an appropriately defined local mean
are modeled, or spatial filtering [large-eddy simulation—(LES)] in which only the
influences of turbulent fluctuations at scales smaller than a prescribed lower limit
(the filter scale) are modeled. Probability density functions (PDFs) are introduced
at this point, as PDF methods are a particularly effective approach for modeling
chemically reacting turbulent flows with radiative heat transfer, and PDF methods
are used in many of the examples that are discussed in subsequent chapters. Specific
models are introduced in Sect. 2.3, and the general notion of turbulence–chemistry
interactions (TCI) is discussed in Sect. 2.4. Finally, extensions to accommodate
multiphase systems are discussed in Sect. 2.5. A radiation source term is retained
in the energy (enthalpy) equation that is presented here, and some key radiation
quantities are defined (e.g., radiant fraction), but detailed discussions of radiation
modeling and turbulence–radiation interaction (TRI) are deferred to Chap. 3.

© The Author(s) 2016
M.F. Modest, D.C. Haworth, *Radiative Heat Transfer in Turbulent Combustion
Systems*, SpringerBriefs in Applied Sciences and Technology,
DOI 10.1007/978-3-319-27291-7_2

2.1   The Unaveraged/Unfiltered Governing Equations (DNS)

The PDEs governing a gas-phase multicomponent reacting system comprising N_S chemical species are expressed here using Cartesian tensor notation. Unless specified otherwise, a Roman index denotes a component of a three-dimensional vector (e.g., $i = 1, 2, 3$), a Greek index denotes a chemical species (e.g., $\alpha = 1, 2, \ldots, N_S$), and the usual summation convention applies over repeated Roman indices within a term:

$$\frac{\partial \rho}{\partial t} + \frac{\partial \rho u_i}{\partial x_i} = 0 \,,$$

$$\frac{\partial \rho u_j}{\partial t} + \frac{\partial \rho u_j u_i}{\partial x_i} = \frac{\partial \tau_{ij}}{\partial x_i} - \frac{\partial p}{\partial x_j} + \rho g_j \ (j = 1, 2, 3) \,,$$

$$\frac{\partial \rho Y_\alpha}{\partial t} + \frac{\partial \rho Y_\alpha u_i}{\partial x_i} = -\frac{\partial J_i^\alpha}{\partial x_i} + S_{\alpha,\text{chem}} \ (\alpha = 1, 2, \ldots, N_S) \,,$$

$$\frac{\partial \rho h}{\partial t} + \frac{\partial \rho h u_i}{\partial x_i} = -\frac{\partial J_i^h}{\partial x_i} + \frac{Dp}{Dt} + \tau_{ij}\frac{\partial u_j}{\partial x_i} + S_{\text{rad}} \,. \qquad (2.1)$$

Here $\mathbf{u}$ denotes the velocity vector, $\mathbf{Y}$ is the vector of mass fractions of the N_S chemical species, and h is the mixture specific enthalpy. Mixture mass density is ρ, pressure is p, body force per unit mass (constant) is $\mathbf{g}$, and $\boldsymbol{\tau}$, $\mathbf{J}^\alpha$, and $\mathbf{J}^h$, denote, respectively, the viscous stress tensor and the molecular flux vectors of species and enthalpy. The chemical production rate for species α, $S_{\alpha,\text{chem}}$, is related to the molar species chemical production rate $\dot{\omega}_\alpha$ and the species molecular weight W_α by $S_{\alpha,\text{chem}} = W_\alpha \dot{\omega}_\alpha$. The volume rate of heating due to radiation (absorption minus emission) is S_{rad}. Additional terms to accommodate multiphase mixtures (soot, liquid fuel sprays, and coal particles) will be introduced in Sect. 2.5. Derivations of Eq. (2.1) can be found in standard combustion textbooks (e.g., [1]). The PDE for h can be replaced with any other suitable form of an energy equation. For low-to-moderate-Mach-number flows, such as those that occur in most combustion devices, it is often convenient to replace the top PDE in Eq. (2.1) (which expresses conservation of mass, or continuity) with a PDE for the pressure; that equation is derived by taking the divergence of the momentum equation [the second PDE in Eq. (2.1)] and invoking the continuity equation to arrive at an elliptic (Poisson) equation for the pressure.

   To complete the description of the physical system, the PDEs are supplemented by thermal ($p = p(\mathbf{Y}, T, \rho)$) and caloric ($T = T(\mathbf{Y}, h, p)$) equations of state, and by specification of fluid transport properties and constitutive relations for stresses and molecular fluxes. Additional properties and equations are needed when radiative heat transfer is considered, and those are introduced in Chap. 3.

For an ideal-gas mixture, the state equations can be written as:

$$p = \rho RT \ (R = R_u/W) \, , \quad h = \sum_{\alpha=1}^{N_S} Y_\alpha \left(\Delta h_{f,\alpha}^0 + \int_{T^0}^{T} c_{p,\alpha}(T')dT' \right) . \tag{2.2}$$

Here R_u is the universal gas constant, W is the mixture molecular weight, $\Delta h_{f,\alpha}^0$ is the enthalpy of formation of species α at reference temperature T^0, and $c_{p,\alpha}$ is the species-α constant-pressure specific heat. The second equation can be used to obtain the temperature, given the species mass fractions, mixture specific enthalpy, and fluid properties. Then the first equation can be used to obtain the pressure, given the density, composition, and temperature; alternatively, in the case where a PDE is solved for the pressure, the first equation can be used to obtain the density, given the pressure, composition, and temperature.

The specific forms of the molecular transport terms (τ_{ij}, J_i^α, and J_i^h) are not central to our discussion of turbulent flames, but for concreteness, one can consider that standard formulations have been adopted (e.g., [2]). For example, the viscous stress normally is written in a form appropriate for a Newtonian fluid, and species and enthalpy transport usually are modeled using multicomponent forms of Fick's and Fourier's laws, respectively.

An arbitrary elementary chemical reaction mechanism involving the N_S species can be written as a set of L reversible reactions:

$$\sum_{\alpha=1}^{N_S} v'_{l\alpha} M_\alpha \rightleftharpoons \sum_{\alpha=1}^{N_S} v''_{l\alpha} M_\alpha \ (l = 1, 2, \dots, L) . \tag{2.3}$$

Here M_α denotes a chemical species symbol, and $v'_{l\alpha}$ and $v''_{l\alpha}$ are the stoichiometric coefficients. Two- and three-body reactions (up to three nonzero values of $v'_{l\alpha}$ and $v''_{l\alpha}$ per reaction) are of most interest for combustion. The corresponding molar rate of production of species α is given by the law of mass action:

$$\dot{\omega}_\alpha = \sum_{l=1}^{L} \left\{ (v''_{l\alpha} - v'_{l\alpha}) \left[k_{l,f}(T) \prod_{\beta=1}^{N_S} c_\beta^{v'_{l\beta}} - k_{l,r}(T) \prod_{\beta=1}^{N_S} c_\beta^{v''_{l\beta}} \right] \right\} , \tag{2.4}$$

where $k_{l,f}(T)$ and $k_{l,r}(T)$ are, respectively, the forward and reverse rate coefficients for the lth reaction (the two are related through an equilibrium constant [1, 3]) and c_β denotes the molar concentration of species β. The rate coefficients usually are expressed in modified Arrhenius form as,

$$k_{l,f}(T) = A_{l,f} T^{b_{l,f}} \exp\{-E_{A\,l,f}/(R_u T)\} , \tag{2.5}$$

where the three coefficients for each reaction (pre-exponential $A_{l,f}$, temperature exponent $b_{l,f}$, and activation energy $E_{A\,l,f}$) may be determined from first principles (in the case of dilute binary elementary reactions) or empirically. An explicit pressure dependence sometimes is included in Eq. (2.5), as well.

Detailed chemical mechanisms (that include all species and reactions for a "complete" representation at the molecular level) and skeletal chemical mechanisms (obtained through systematic elimination of unimportant species and/or reactions starting from a detailed mechanism) generally contain only elementary reactions and can be expressed in the form of Eq. (2.3) through Eq. (2.5). Reduced chemical mechanisms (obtained by further systematic reduction or empirically), on the other hand, may contain nonelementary reactions and/or algebraic constraints, so that Eq. (2.4) does *not* follow from Eq. (2.3); the modified Arrhenius form still generally is used for the reaction rates.

In the remainder of this monograph, it is assumed that an appropriate detailed, skeletal, or reduced chemical mechanism is available as needed to provide the chemical source terms in the unaveraged/unfiltered species mass fraction Eq. (2.1) as functions of the local species mass fractions, temperature, and pressure: $\mathbf{S}_{\text{chem}} = \mathbf{S}_{\text{chem}}(\mathbf{Y}, T, p)$. Examples of DNS where the unaveraged/unfiltered equations are solved numerically are provided in Chap. 5, but most of the results that are discussed in this monograph are from RANS and LES modeling studies, which are discussed in the following section.

For subsequent discussions of radiative heat transfer in combustion, it is convenient to introduce the notion of a "radiant fraction." This is the ratio of the net radiative energy that is emitted by a flame externally to its surroundings to the chemical energy that is supplied to the flame in the reactants. To make quantitative comparisons between experiments and modeling studies, it is important that both the numerator (net radiative energy emitted) and the denominator (chemical energy of the reactants) be defined consistently. Standard practice is to quantify the reactant chemical energy using the lower heating value of the fuel (denoted Δh_c, with units of energy per unit mass of fuel burned); this is a quantity that can be found in standard thermodynamic tables (e.g., [3]). The rate at which chemical energy is supplied to the flame is then $\dot{m}_F \Delta h_c$, where $\dot{m}_F$ is the mass flow rate of fuel into the flame or combustor. The rate of radiative heat transfer from the flame to the surroundings, $\dot{Q}_{\text{rad}}$, can be defined formally as the integral of the radiation source term in the enthalpy equation over a control volume that contains the flame: $\dot{Q}_{\text{rad}} \equiv -\int_V S_{\text{rad}} dV = \int_S \mathbf{q}_{\text{rad}} \cdot \mathbf{n} dS$, where $\mathbf{q}_{\text{rad}}$ is the radiant heat flux vector ($S_{\text{rad}} = -\nabla \cdot \mathbf{q}_{\text{rad}}$), and $\mathbf{n}$ is the local unit outward-pointing normal vector to the surface S of the control volume V. In terms of these two quantities, the radiant fraction χ_R is:

$$\chi_R \equiv \frac{\dot{Q}_{\text{rad}}}{\dot{m}_F \Delta h_c} \,. \tag{2.6}$$

The value of χ_R thus depends on the control volume that is used, and in comparing numerical simulation results with experimental results, it is important to use consistent definitions.

2.2 Reynolds Averaging (RANS) and Spatial Filtering (LES)

As noted earlier, it is not feasible to resolve all scales of interest in numerical simulations of practical turbulent combustion systems, and one of two modeling frameworks is usually adopted to address this. Reynolds averaging (RANS) is discussed first. Then spatial filtering (LES) is discussed in somewhat less detail, focusing mainly on its similarities and differences with respect to RANS. Several examples from RANS and LES modeling studies are presented in subsequent chapters.

2.2.1 Reynolds Averaging (RANS)

In a turbulent flame or combustion device, estimates of the local mean value of a physical quantity of interest (e.g., velocity, species mass fractions, temperature, etc.) can be obtained by long-time averaging in a statistically stationary flow (e.g., measuring time series of the species mass fractions at a fixed radial and axial position in an axisymmetric turbulent jet flame [4]), by spatial averaging in configurations having one or more directions of spatially homogeneity (e.g., averaging parallel to the rod in a ducted rod-stabilized flame [5]), or by ensemble- or phase-averaging (e.g., measuring the in-cylinder velocity at a specified location for a given piston position over many engine cycles in a reciprocating-piston internal combustion— IC—engine [6]). This approach to reducing the dynamic range of scales that must be resolved computationally is the basis for Reynolds-averaged modeling studies (RANS) of turbulent reacting flows. As defined here, RANS is not limited to statistically stationary systems; the designation "URANS" (unsteady RANS) is sometimes used for situations where RANS is applied to systems that are not statistically stationary, but the distinction is not necessary for present purposes.

Equations for mean quantities (denoted using angled brackets $\langle \ \rangle$) can be derived by taking the mean of Eq. (2.1). For this purpose, let $Q = Q(\mathbf{x}, t)$ denote a random variable that corresponds to any of the physical quantities of interest in turbulent combustion: velocity, species mass fractions, temperature, etc. The instantaneous value of Q can be decomposed into the sum of its local mean value and a fluctuation with respect to that mean (denoted using a prime). An appropriate mean is defined as described in the previous paragraph for the system of interest, such that the $\langle \ \rangle$ operator commutes with spatial and/or temporal differentiation, and the mean and the fluctuating components have the following properties:

$$Q(\mathbf{x}, t) = \langle Q(\mathbf{x}, t)\rangle + Q'(\mathbf{x}, t) \,, \ \langle Q'(\mathbf{x}, t)\rangle = 0 \,, \ \langle\langle Q(\mathbf{x}, t)\rangle\rangle = \langle Q(\mathbf{x}, t)\rangle \,. \quad (2.7)$$

This is a *Reynolds decomposition*. For variable-density flows, it is advantageous to work with density-weighted (Favre-averaged) mean quantities, as the resulting equations have a simpler form that is similar to that of the original unaveraged/unfiltered equations. We introduce the density-weighted mean of Q, denoted $\tilde{Q}$, as:

$$\tilde{Q} \equiv \langle \rho Q \rangle / \langle \rho \rangle \ . \tag{2.8}$$

The instantaneous value of Q also can be decomposed into the sum of its local Favre-mean value and a fluctuation with respect to that mean (denoted using a double prime):

$$Q(\mathbf{x}, t) = \tilde{Q}(\mathbf{x}, t) + Q''(\mathbf{x}, t) \ , \ \ \widetilde{Q''}(\mathbf{x}, t) = 0 \ , \ \ \tilde{\tilde{Q}}(\mathbf{x}, t) = \tilde{Q}(\mathbf{x}, t) \ . \tag{2.9}$$

With these definitions and decompositions, the PDEs for mean quantities corresponding to Eq. (2.1) can be written as:

$$\frac{\partial \langle \rho \rangle}{\partial t} + \frac{\partial \langle \rho \rangle \tilde{u}_i}{\partial x_i} = 0 \ ,$$

$$\frac{\partial \langle \rho \rangle \tilde{u}_j}{\partial t} + \frac{\partial \langle \rho \rangle \tilde{u}_j \tilde{u}_i}{\partial x_i} = \frac{\partial (\langle \rho \rangle \tilde{u}_j \tilde{u}_i - \langle \rho \rangle \widetilde{u_j u_i})}{\partial x_i}$$

$$+ \frac{\partial \langle \tau_{ij} \rangle}{\partial x_i} - \frac{\partial \langle p \rangle}{\partial x_j} + \langle \rho \rangle g_j \ \ (j = 1, 2, 3) \ ,$$

$$\frac{\partial \langle \rho \rangle \widetilde{Y}_\alpha}{\partial t} + \frac{\partial \langle \rho \rangle \widetilde{Y}_\alpha \tilde{u}_i}{\partial x_i} = \frac{\partial (\langle \rho \rangle \widetilde{Y}_\alpha \tilde{u}_i - \langle \rho \rangle \widetilde{Y_\alpha u_i})}{\partial x_i}$$

$$- \frac{\partial \langle J_i^\alpha \rangle}{\partial x_i} + \langle S_{\alpha,\mathrm{chem}} \rangle \ \ (\alpha = 1, 2, \dots, N_S) \ ,$$

$$\frac{\partial \langle \rho \rangle \tilde{h}}{\partial t} + \frac{\partial \langle \rho \rangle \tilde{h} \tilde{u}_i}{\partial x_i} = \frac{\partial (\langle \rho \rangle \tilde{h} \tilde{u}_i - \langle \rho \rangle \widetilde{h u_i})}{\partial x_i}$$

$$- \frac{\partial \langle J_i^h \rangle}{\partial x_i} + \frac{D \langle p \rangle}{Dt} + \Phi + \langle S_{rad} \rangle \ , \tag{2.10}$$

where Φ is the mean viscous dissipation rate of kinetic energy to heat,

$$\Phi \equiv \langle \tau_{ij} \frac{\partial u_j}{\partial x_i} \rangle \ . \tag{2.11}$$

The form of Eq. (2.10) is similar to that of Eq. (2.1), with the exception of the first term on the right-hand side of the mean momentum, species, and enthalpy equations. The quantity in parentheses in each of those terms can be written as:

$$\langle \rho \rangle \tilde{u}_j \tilde{u}_i - \langle \rho \rangle \widetilde{u_j u_i} = -\langle \rho \rangle \widetilde{u_j'' u_i''} \equiv \tau_{T\,ij} \ ,$$

$$\langle \rho \rangle \widetilde{Y}_\alpha \tilde{u}_i - \langle \rho \rangle \widetilde{Y_\alpha u_i} = -\langle \rho \rangle \widetilde{Y_\alpha'' u_i''} \ ,$$

$$\langle \rho \rangle \tilde{h} \tilde{u}_i - \langle \rho \rangle \widetilde{h u_i} = -\langle \rho \rangle \widetilde{h'' u_i''} \ . \tag{2.12}$$

These represent turbulent fluxes of momentum, species, and enthalpy, respectively. The first of these is the apparent turbulent stress, τ_{Tij}. These turbulent fluxes, and terms in Eq. (2.10) that correspond to the means of any other nonlinear terms in Eq. (2.1), are not closed at the level of mean quantities. In particular, for a given chemical mechanism $S_{chem} = S_{chem}(Y, T, p)$:

$$\langle S_{chem}(Y, T, p) \rangle = \langle S_{chem}(x, t) \rangle \neq S_{chem}(\langle Y(x, t) \rangle, \langle T(x, t) \rangle, \langle p(x, t) \rangle) , \qquad (2.13)$$

and the differences are manifestations of turbulence–chemistry interactions in the RANS context.

More formally, one can introduce the *probabilistic mean* or *expected value* of a random variable and its associated *probability density function* (PDF) [7–9]. Then the mean of Q, denoted $\langle Q \rangle$, can be expressed in terms of the PDF of Q, $f_Q(\psi; x, t)$: $\langle Q \rangle = \langle Q(x, t) \rangle = \int_{-\infty}^{\infty} \psi f_Q(\psi; x, t) d\psi$. The PDF quantifies the probability of Q taking on a particular value: $f_Q(\psi; x, t) d\psi = \mathrm{Prob}\{\psi \leq Q(x, t) < \psi + d\psi\}$. In turbulent combustion, a central role is played by the joint PDF of the composition variables $\boldsymbol{\phi}(x, t)$ that define the local thermochemical state of the reacting system [8]. For most of the systems considered here, a suitable set of composition variables is the species mass fractions Y and the mixture specific enthalpy h [Eq. (2.2)]: $\{\boldsymbol{\phi}\} = \{Y, h\}$. Further simplifications are possible in some cases: e.g., in systems where a mixture fraction and/or small set of reaction progress variables suffices to describe the thermochemical state. We then introduce the *composition joint PDF*, $f_{\boldsymbol{\phi}}(\boldsymbol{\psi}; x, t)$; this is the one-point, one-time Eulerian joint PDF of the event $\{\boldsymbol{\phi}(x, t) = \boldsymbol{\psi}\}$. Properties of this PDF include: $f_{\boldsymbol{\phi}}(\boldsymbol{\psi}; x, t) \geq 0$; and $\int f_{\boldsymbol{\phi}}(\boldsymbol{\psi}; x, t) d\boldsymbol{\psi} = 1$. Here and in the following, integration over the entire sample space is implied, unless specified otherwise. The most important property of the PDF for the present discussion is that the mean of any function of $\boldsymbol{\phi}(x, t)$, $Q = Q(\boldsymbol{\phi})$, can be expressed as an integral over the PDF:

$$\langle Q \rangle = \langle Q(x, t) \rangle = \int Q(\boldsymbol{\psi}) f_{\boldsymbol{\phi}}(\boldsymbol{\psi}; x, t) d\boldsymbol{\psi}. \qquad (2.14)$$

Again, for variable-density flows, it is advantageous to work with density-weighted (Favre-averaged) quantities. The *Favre PDF*, $\tilde{f}_{\boldsymbol{\phi}}, (\boldsymbol{\psi}; x, t)$ and the *conventional PDF*, $f_{\boldsymbol{\phi}}(\boldsymbol{\psi}; x, t)$, are related by:

$$\langle \rho(x, t) \rangle \tilde{f}_{\boldsymbol{\phi}}(\boldsymbol{\psi}; x, t) = \rho(\boldsymbol{\psi}) f_{\boldsymbol{\phi}}(\boldsymbol{\psi}; x, t) , \qquad (2.15)$$

where $\langle \rho \rangle$ is the mean mixture mass density. Conventional ($\langle \ \rangle$) and density-weighted ($\tilde{\ }$) means of any function of $\boldsymbol{\phi}$ are readily computed from the PDF. For any $Q = Q(\boldsymbol{\phi})$ we have,

$$\int Q(\boldsymbol{\psi}) f_{\boldsymbol{\phi}}(\boldsymbol{\psi}; x, t) d\boldsymbol{\psi} = \langle \rho(x, t) \rangle \int Q(\boldsymbol{\psi}) \rho^{-1}(\boldsymbol{\psi}) \tilde{f}_{\boldsymbol{\phi}}(\boldsymbol{\psi}; x, t) d\boldsymbol{\psi} = \langle Q(x, t) \rangle ,$$

$$\int Q(\boldsymbol{\psi})\tilde{f}_\phi(\boldsymbol{\psi};\mathbf{x},t)d\boldsymbol{\psi} = \langle\rho(\mathbf{x},t)\rangle^{-1}\int Q(\boldsymbol{\psi})\rho(\boldsymbol{\psi})f_\phi(\boldsymbol{\psi};\mathbf{x},t)d\boldsymbol{\psi} = \tilde{Q}(\mathbf{x},t).$$

$$(2.16)$$

All of the earlier properties of conventional and Favre-averaged mean quantities follow immediately. In terms of $f_\phi(\boldsymbol{\psi};\mathbf{x},t)$, the mean chemical source term in the RANS species equation can be written as:

$$\langle S_{\alpha,\mathrm{chem}}(\mathbf{x},t)\rangle = \int S_{\alpha.\mathrm{chem}}(\boldsymbol{\psi})f_\phi(\boldsymbol{\psi};\mathbf{x},t)d\boldsymbol{\psi}\ .$$

$$(2.17)$$

That is, with knowledge of the composition joint PDF, the mean chemical source term can be calculated without further approximation (for a given chemical mechanism). This shifts the modeling effort from that of expressing $\langle \mathbf{S}_{\mathrm{chem}}\rangle$ in terms of other mean quantities to that of computing the composition joint PDF $f_\phi(\boldsymbol{\psi};\mathbf{x},t)$. A PDF governing the evolution of $f_\phi(\boldsymbol{\psi};\mathbf{x},t)$ or of $\tilde{f}_\phi(\boldsymbol{\psi};\mathbf{x},t)$ can be derived from Eq. (2.1), and this is discussed in Sect. 2.3 below.

2.2.2 Spatial Filtering (LES)

The local spatially filtered value of a physical quantity Q, denoted $\langle Q\rangle_\Delta$, is defined as [10–12],

$$\langle Q\rangle_\Delta = \langle Q(\mathbf{x},t)\rangle_\Delta \equiv \int Q(\mathbf{y},t)G(|\mathbf{x}-\mathbf{y}|)d\mathbf{y}\ ,$$

$$(2.18)$$

where integration is over the entire flow domain. Here the low-pass spatial filter function, $G(|\mathbf{x}-\mathbf{y}|)$, satisfies $\int G(\mathbf{x})d\mathbf{x} = 1$, has a characteristic filter width Δ, is independent of spatial location, and is isotropic. The latter two properties often are not satisfied by the filters that are used in practice, but simplify the exposition. The spatially filtered values $\langle Q\rangle_\Delta$ thus depend on the filter function G and on the filter size Δ that are selected. Properties of "well-behaved" filter functions for LES are discussed in [11] and in Chap. 13 of [13], for example. In practice, the filter size Δ is usually taken to be equal to, or at least proportional to, the local CFD mesh size.

  The instantaneous value of any physical quantity $Q = Q(\mathbf{x},t)$ in a turbulent flow then can be decomposed into a filtered component (the resolved field) and a fluctuation about the filtered component (the residual field): $Q(\mathbf{x},t) = \langle Q(\mathbf{x},t)\rangle_\Delta + Q'(\mathbf{x},t)$. In contrast to a Reynolds decomposition, the filtering operation does not commute with spatial and temporal differentiation (although this is often ignored, in practice), and the filtered value of the fluctuating component is not equal to zero [see Eq. (2.7)]:

$$Q(\mathbf{x},t) = \langle Q(\mathbf{x},t)\rangle_\Delta + Q'(\mathbf{x},t)\ ,\ \langle Q'(\mathbf{x},t)\rangle_\Delta \neq 0\ ,\ \langle\langle Q(\mathbf{x},t)\rangle_\Delta\rangle_\Delta \neq \langle Q(\mathbf{x},t)\rangle_\Delta\ .$$

$$(2.19)$$

As was the case in RANS, density weighting is useful in LES of variable-density flows:

$$\widehat{Q} \equiv \langle \rho Q \rangle_\Delta / \langle \rho \rangle_\Delta = \int \rho(\mathbf{y}, t) Q(\mathbf{y}, t) G(|\mathbf{x} - \mathbf{y}|) d\mathbf{y} / \langle \rho \rangle_\Delta , \qquad (2.20)$$

and the corresponding fluctuation is denoted using a double prime: $Q(\mathbf{x}, t) = \widehat{Q}(\mathbf{x}, t) + Q''(\mathbf{x}, t)$.

A conventional PDF ($f_{\Delta,\phi}$) or density-weighted PDF ($\hat{f}_{\Delta,\phi}$) of subfilter-scale fluctuations (a subfilter-scale PDF) can be introduced such that:

$$\int Q(\boldsymbol{\psi}) f_{\Delta,\phi}(\boldsymbol{\psi}; \mathbf{x}, t) d\boldsymbol{\psi} = \langle Q(\mathbf{x}, t) \rangle_\Delta ,$$

$$\int Q(\boldsymbol{\psi}) \hat{f}_{\Delta,\phi}(\boldsymbol{\psi}; \mathbf{x}, t) d\boldsymbol{\psi} = \widehat{Q}(\mathbf{x}, t) . \qquad (2.21)$$

In the literature, the term "filtered density function" (FDF) was used in early work on PDF-based subfilter-scale modeling for LES, but "PDF" is used in most of the more recent literature, for reasons that are explained in [9]. Here we will refer to LES using a PDF-based subfilter-scale modeling approach as LES/PDF, to distinguish it from RANS/PDF. The interpretation of the prime or double-prime notation for fluctuating quantities (fluctuation with respect to the local mean value versus fluctuation with respect to the local spatially filtered value) will be clear from the context (RANS versus LES).

In LES, one formally applies a low-pass spatial filter with a characteristic filter size Δ to Eq. (2.1) to arrive at the PDEs that govern the evolution of the larger (resolved) scales (those larger than Δ), and the effects of smaller scales (smaller than Δ) on the resolved fields are represented by introducing subfilter-scale models. The PDEs for filtered quantities have a form that is similar to Eq. (2.10), but there are additional terms that correspond to the nonzero values of the filtered fluctuations in LES [Eq. (2.19)]. As was the case for the mean chemical source term in RANS, the filtered chemical source term in LES does not have the same functional form in terms of spatially filtered quantities as the chemical source term in the unaveraged/unfiltered species equations in terms of local instantaneous quantities [see Eq. (2.13)]:

$$\langle \mathbf{S}_{\text{chem}}(\mathbf{Y}, T, p) \rangle_\Delta = \langle \mathbf{S}_{\text{chem}}(\mathbf{x}, t) \rangle_\Delta \neq \mathbf{S}_{\text{chem}}(\langle \mathbf{Y}(\mathbf{x}, t) \rangle_\Delta, \langle T(\mathbf{x}, t) \rangle_\Delta, \langle p(\mathbf{x}, t) \rangle_\Delta) , \qquad (2.22)$$

and the differences are manifestations of turbulence–chemistry interactions in the LES context. On the other hand, the filtered chemical source term in LES can be written in terms of the subfilter-scale composition PDF as [see Eq. (2.17)]:

$$\langle S_{\alpha,\text{chem}}(\mathbf{x}, t) \rangle_\Delta = \int S_{\alpha.\text{chem}}(\boldsymbol{\psi}) f_{\Delta,\phi}(\boldsymbol{\psi}; \mathbf{x}, t) d\boldsymbol{\psi} . \qquad (2.23)$$

Similarities and differences between RANS/PDF and LES/PDF are discussed further in Sects. 2.3 and 2.4.

2.3   Turbulence and Turbulent Combustion Modeling

Because PDF methods are particularly useful and effective in accounting for
effects of unresolved turbulent fluctuations in both RANS and in LES, the emphasis
here is on approaches based on modeling and solving a transport equation for the
joint PDF of a set of composition variables that is sufficient to define the local
thermochemical state of the reacting system: that is, a set of variables $\boldsymbol{\phi}$ from
which one can determine the mass density ρ, chemical source terms $\mathbf{S}_{\mathrm{chem}}$, and
any other fluid properties that are needed in the unaveraged/unfiltered equations
(e.g., radiative properties—see Chap. 3). For most of the systems that are considered
here, mass fractions $\mathbf{Y}$ of the N_S chemical species in the system plus mixture specific
enthalpy h are an appropriate set, and in that case, the number of composition
variables is $N_S + 1$. As in the previous section, we begin with an overview of
modeling in the RANS context. Then LES modeling is described in less detail,
focusing as before on similarities and differences with respect to RANS.

2.3.1   RANS Modeling

A PDE governing the evolution of the composition joint PDF can be derived from
Eq. (2.1) [7, 8]. To better focus on the processes/terms that are of primary interest
for chemically reacting turbulent flows with radiative heat transfer, here we consider
a constant-pressure combustion system where the viscous dissipation term in the
enthalpy equation is negligible, and simplify the enthalpy equation accordingly.
These are reasonable approximations for many practical combustion systems, and
can be relaxed when necessary. Then in terms of the conventional PDF, the
composition PDF transport equation can be written as:

$$\frac{\partial \rho f_{\phi}}{\partial t} + \frac{\partial \rho \tilde{u}_i f_{\phi}}{\partial x_i} + \frac{\partial S_{\alpha,\mathrm{chem}} f_{\phi}}{\partial \psi_{\alpha}} = -\frac{\partial}{\partial x_i}[\langle u_i'' | \boldsymbol{\psi} \rangle \rho f_{\phi}] + \frac{\partial}{\partial \psi_{\alpha}}[\langle \frac{\partial J_i^{\alpha}}{\partial x_i} | \boldsymbol{\psi} \rangle f_{\phi}]$$

$$-\delta_{\alpha(h)}\frac{\partial}{\partial \psi_{\alpha}}[\langle S_{\mathrm{rad}} | \boldsymbol{\psi} \rangle f_{\phi}] . \qquad (2.24)$$

Here $\rho = \rho(\boldsymbol{\psi})$, $\alpha = 1,\ldots,N_S + 1$, where the first N_S values correspond to the species mass fractions, the last value corresponds to enthalpy
(so that $S_{N_S+1,\mathrm{chem}} = 0$), and summation is implied over repeated values of both
i (Cartesian coordinate index) and α (composition variable index) within a term.
The notation $\delta_{\alpha(h)}$ denotes the Kronecker delta function, and is equal to zero for
all values of α except for the one corresponding to enthalpy ($\alpha = N_S + 1$).
The notation $\langle Q | \boldsymbol{\psi} \rangle$ in each term on the right-hand side denotes the conditional
expectation of quantity Q, conditioned on the event $\boldsymbol{\phi} = \boldsymbol{\psi}$. For cases where Q is
neither uniquely determined by $\boldsymbol{\phi}$ nor is statistically independent of $\boldsymbol{\phi}$ (which is the
case for the three terms shown), these are unclosed terms that require modeling.

In the composition PDF transport equation, the terms corresponding to advection by the mean velocity and chemical reaction appear in closed form (left-hand side; no further modeling is required), while the terms corresponding to transport by turbulent velocity fluctuations u_i'' (turbulent diffusion, or macro-mixing), molecular transport J_i^α (molecular mixing, or micro-mixing), and radiation (see Chap. 3) on the right-hand side must be modeled. The (still unclosed) PDEs for mean quantities [Eq. (2.10)] and higher-order statistics can be derived from the composition PDF equation using Eqs. (2.7) and (2.16) and the properties of the PDF [8].

One-point, one-time statistics (and hence the PDFs considered here) provide no explicit information on spatial structure, length scales, or time scales in turbulent reacting flows. Both the formal axiomatic development and the methods that are used to estimate mean values in physical systems (e.g., time averaging, spatial averaging, and/or ensemble averaging) suggest that mean quantities and PDFs in general should evolve on length and time scales that are not smaller than the largest turbulence scales (the integral scales). Therefore, the spatial derivatives and temporal derivatives that appear in PDF transport equations and in any equations that are derived from them (e.g., moment equations) are expected to be smooth at scales that are of the order of the integral scales of the turbulent flow.

As an example, the simplest model for turbulent transport is a gradient transport approximation, where all composition variables diffuse down their local mean gradients with an apparent turbulent diffusivity $\Gamma_{T\phi}$, and the simplest model for molecular transport is a mixing model where all composition variables relax toward the local mean composition on a turbulent time scale $\tau = 1/\omega$, where ω is a "turbulent frequency" (interaction by exchange with the mean—IEM). In that case, the modeled (with the exception of the radiative source term—see Chap. 3) composition PDF transport equation is,

$$\frac{\partial \rho f_\phi}{\partial t} + \frac{\partial \rho \tilde{u}_i f_\phi}{\partial x_i} + \frac{\partial S_\alpha f_\phi}{\partial \psi_\alpha} = \frac{\partial}{\partial x_i}\left[\Gamma_{T\phi}\frac{\partial \rho f_\phi/\langle\rho\rangle}{\partial x_i}\right] + \frac{1}{2}\frac{\partial}{\partial \psi_\alpha}[C_\phi \omega(\psi_\alpha - \tilde{\phi}_\alpha)\rho f_\phi]$$

$$- \delta_{\alpha(h)}\frac{\partial}{\partial \psi_\alpha}[\langle S_{rad}|\boldsymbol{\psi}\rangle f_\phi] , \tag{2.25}$$

where C_ϕ is a model constant for the mixing model. The apparent turbulent diffusivity can be written as $\Gamma_{T\phi} = \mu_T/\sigma_\phi$, where μ_T is the apparent turbulent viscosity and σ_ϕ is an apparent turbulent Schmidt or Prandtl number. The mean velocity field and turbulence scales needed to form μ_T and ω must be provided separately. This can be done by solving the modeled mean momentum equation [Eq. (2.10)] using an appropriate turbulence model for the apparent turbulent stress τ_{Tij}. For example, in a standard two-equation $k-\varepsilon$ turbulence model, μ_T is given by:

$$\mu_T = C_\mu \langle\rho\rangle k^2/\varepsilon , \tag{2.26}$$

and τ_{Tij} is given by:

$$\tau_{Tij} \equiv -\langle\rho\rangle\widetilde{u_i''u_j''} = \mu_T(\frac{\partial\tilde{u}_i}{\partial x_j} + \frac{\partial\tilde{u}_j}{\partial x_i}) - \frac{2}{3}\mu_T\frac{\partial\tilde{u}_l}{\partial x_l}\delta_{ij} - \frac{2}{3}\langle\rho\rangle k\delta_{ij} \ . \tag{2.27}$$

Here C_μ is a model constant, and modeled transport equations are solved for the turbulence kinetic energy k and for the viscous rate of dissipation of turbulence kinetic energy ε. The turbulence frequency is then $\omega = \tau^{-1} = \varepsilon/k$. Any other turbulence model can be used as an alternative to $k - \varepsilon$, and appropriate expressions for $\Gamma_{T\phi}$ and ω must be provided. Closed (with the exception of the radiation term in the enthalpy equation) PDEs for mean quantities [Eq. (2.10)] and higher-order statistics can be derived from the modeled composition PDF equation using Eqs. (2.7) and (2.16) [8].

The composition PDF equation is an integral-differential equation in up to $N_S + 5$ independent variables: $N_S + 1$ composition-space coordinates, plus up to three spatial coordinates and time. The equation has an integral character because mean quantities that appear in the equation ($\langle\rho\rangle$, $\tilde{\phi}_\alpha$) can be expressed as integrals over the composition PDF. Conventional Eulerian grid-based methods (e.g., finite-difference, finite-volume, or finite-element methods) are not practicable for such high-dimensional systems, and stochastic (Monte Carlo) methods have been developed as an alternative. The prevailing approach is a stochastic Lagrangian particle Monte Carlo method [7, 8]. In that case, a system of notional computational particles is introduced to represent the reacting gas mixture, and an equivalent stochastic particle system can be constructed corresponding to any realizable modeled PDF transport equation. An equivalent stochastic particle system governing the increments in the positions and compositions of notional particles over a time increment dt corresponding to Eq. (2.25) is [8]:

$$dx_i^* = \tilde{u}_i^* dt + \left(\langle\rho\rangle^{-1}\frac{\partial\Gamma_{T\phi}}{\partial x_i}\right)^* dt + \left(2\langle\rho\rangle^{-1}\Gamma_{T\phi}\right)^{*1/2} dW_i \ \ (i = 1, 2, 3) \ ,$$

$$d\phi_\alpha^* = \rho^{-1}(\boldsymbol{\phi}^*)S_{\alpha,\text{chem}}(\boldsymbol{\phi}^*)dt - \frac{1}{2}C_\phi(\phi_\alpha^* - \tilde{\phi}_\alpha)\omega dt$$

$$+\delta_{\alpha(h)}\rho^{-1}(\boldsymbol{\phi}^*)\langle S_{\text{rad}}|\boldsymbol{\phi}^*\rangle dt \ \ (\alpha = 1, \ldots, N_S + 1) \ . \tag{2.28}$$

Here a superscript $*$ refers to a notional particle value, and $\mathbf{W}(t)$ is an isotropic vector Wiener process. The increment $d\mathbf{W}$ is a joint normal random vector with zero mean and with covariance $dt\delta_{ij}$:

$$\langle dW_i\rangle = 0 \ , \ \langle dW_i dW_j\rangle = dt\delta_{ij} \ . \tag{2.29}$$

In a numerical implementation, for example, dW_i can be discretized (approximated) as $\Delta W_i = W_i(t + \Delta t) - W_i(t) = \eta_i\Delta t^{1/2}$, where $\boldsymbol{\eta}$ is a vector of three independent standardized Gaussian (zero mean, unit variance) random variables and Δt is the computational time step.

In practice, the particle method is implemented in a conventional finite-volume Eulerian CFD solver, where there are 20–30 particles (say) per finite-volume cell. The coupled mean continuity (or mean pressure), mean momentum, and turbulence model equations are solved using the finite-volume solver. Each particle moves in physical space according to the local mean velocity (interpolated from the finite-volume solver to the particle position), plus a random component to account for transport by turbulent velocity fluctuations; each particle's composition evolves according to the specified chemical mechanism, plus a contribution to account for molecular transport (mixing), and contributions to account for radiative heat transfer [Eq. (2.28)]. The turbulence scales needed to form $\Gamma_{T,\phi}$ and ω are taken from the turbulence model that is used in the finite-volume calculation. In the case of a pressure-based finite-volume solver (the usual approach for low-Mach-number systems), the principal quantity that is fed to the finite-volume solver from the particles is the mean density, which (in the simplest case) is calculated as the average over the particle values in each finite-volume cell (Fig. 2.1). Important issues including consistency, convergence, and statistic error are discussed in [8].

This can be thought of (somewhat simplistically) as a particle-in-cell method to account for the influences of unresolved turbulent fluctuations on the local mean fields. For example, the change in mean composition for a finite-volume cell due to chemical reactions over a computational time step is simply the (appropriately weighted) average of the changes for the particles in that cell. If the change in

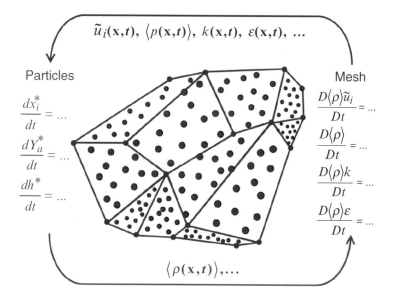

Fig. 2.1 A schematic illustration of a hybrid Lagrangian particle/Eulerian mesh composition PDF method. The figure shows a two-dimensional slice through an arbitrary three-dimensional unstructured finite-volume mesh. *Blue lines* represent finite-volume cell edges, and *red circles* represent notional PDF particles [8]

cell-mean composition due to chemical reactions were to be computed instead based
on cell-mean values of composition and temperature (which are, by definition, the
appropriately weighted averages over the values for the particles in that cell) rather
than on the particle values themselves (a locally well-stirred-reactor—WSR—
model), the result would be different, and the differences are manifestations of
turbulence–chemistry interactions (Sect. 2.4).

2.3.2 LES Modeling

There are important conceptual differences between RANS/PDF and LES/PDF. For
example, the subfilter-scale PDF varies in time even for statistically stationary flows,
and in all three spatial coordinates even for flows having one or more directions
of statistical homogeneity. And the subfilter-scale PDF is itself a random quantity,
in contrast to the PDF that is used in RANS. However, in practice, essentially
similar approaches are followed for RANS/PDF and LES/PDF. The subfilter-scale
composition PDF equation can be derived from Eq. (2.1) using the definitions given
in Sect. 2.2.2, and the equation has a form that is similar to that of Eq. (2.24) [8].
Closure models to account for effects of unresolved turbulent velocity fluctuations
with respect to the local spatially filtered velocity and for effects of molecular
transport are invoked that are similar to those used for the RANS/PDF equation
[Eq. (2.25)]. And a notional particle description is used that is similar to that used for
RANS/PDF [Eq. (2.28)]. The key differences between RANS/PDF and LES/PDF at
the equation level are in the physical interpretations of the resolved fields that appear
(*mean* quantities in RANS, versus *spatially filtered* quantities in LES) and of the
corresponding unresolved turbulent fluctuations (*all fluctuations* with respect to the
local mean values in RANS, versus only the *subfilter-scale fluctuations* with respect
to the local spatially filtered values in LES), and these differences are reflected in
the modeling.

For example, the particle equations for LES/PDF are essentially the same as those
for RANS/PDF, but with the conventional and density-weighted mean quantities
in Eq. (2.28) replaced by the corresponding conventional and density-weighted
spatially filtered values. The LES/PDF particle equations can be written as:

$$dx_i^* = \hat{u}_i^* dt + \left(\langle \rho \rangle_\Delta^{-1} \frac{\partial \Gamma_{T\Delta\phi}}{\partial x_i} \right)^* dt + \left(2\langle \rho \rangle_\Delta^{-1} \Gamma_{T\Delta\phi} \right)^{*\,1/2} dW_i \quad (i = 1, 2, 3) \,,$$

$$d\phi_\alpha^* = \rho^{-1}(\boldsymbol{\phi}^*) S_{\alpha,\text{chem}}(\boldsymbol{\phi}^*) dt - \frac{1}{2} C_{\phi\Delta} (\phi_\alpha^* - \hat{\phi}_\alpha) \omega_\Delta dt$$

$$+ \delta_{\alpha(h)} \rho^{-1}(\boldsymbol{\phi}^*) \langle S_{\text{rad}} | \boldsymbol{\phi}^* \rangle_\Delta dt \quad (\alpha = 1, \dots, N_S + 1) \,. \qquad (2.30)$$

In LES, the subfilter-scale PDF varies on length scales down to the filter scale Δ,
versus the turbulence integral length scale l_T in the case of RANS/PDF; normally
$\Delta < l_T$. The apparent turbulent diffusivity $\Gamma_{T\Delta\phi}$ in LES accounts only for the

influences of subfilter-scale velocity fluctuations, and is generally smaller than the apparent turbulent diffusivity $\Gamma_{T\phi}$ in RANS. Similarly, the appropriate mixing frequency ω_Δ for LES is one that corresponds to subfilter-scale fluctuations only, so that ω_Δ is generally larger that the corresponding RANS mixing frequency ω, and the values of the mixing model constants ($C_{\phi\Delta}$ for LES versus C_ϕ for RANS) are also different. In most cases, the appropriate turbulence scales for subfilter-scale processes in LES (the specific functional forms for $\Gamma_{T\Delta\phi}$ and ω_Δ) are constructed using the filter size Δ as a length scale, and an estimate of the subfilter-scale turbulence kinetic energy k_Δ that is consistent with the subfilter-scale turbulence model that is used in the filtered momentum equation for the LES. For LES/PDF, typically a smaller number of particles per cell is used compared to RANS/PDF, with the reasoning that only the subfilter-scale fluctuations (versus *all* of the fluctuations with respect to the local mean) are represented by the particles. Further details can be found in Chaps. 5 and 6 of [8].

In both RANS/PDF and LES/PDF methods, there are many variants in the specific physical models that are used (e.g., mixing models). There are alternatives to the stochastic Lagrangian particle methods that have been described here for solving modeled PDF transport equations (e.g., stochastic and deterministic Eulerian field methods [8, 14]). And higher-order PDF closures have been developed that do not require as much external information concerning the mean (or filtered) velocity field and turbulence scales (e.g., velocity-composition and velocity-composition-frequency PDF methods [7–9]. It should also be mentioned that there are approaches to turbulent combustion modeling that are not based on solving a modeling PDF transport equation, and some of those will be mentioned in the examples that are provided in subsequent chapters.

2.4 TCI in RANS and LES

Chemical reactions influence fluid flow primarily through changes in the local temperature (conversion of chemical energy to sensible energy), which in turn affect the local mass density and other fluid properties. For constant-pressure adiabatic combustion of stoichiometric hydrocarbon-air reactants at ambient conditions, the temperature changes by approximately a factor of seven due to heat release (from $\sim$300 K to $\sim$2100 K). The flow, in turn, influences the chemistry by (for example) straining or curving the local flame structure (modifying the local balance between molecular transport and heat release) in a manner that changes the local heat-release rate. Generically, these effects are referred to as "turbulence–chemistry interactions" (TCI), and they have been the subject of extensive research in the combustion community over the last several decades. In RANS or in LES, the influences of unresolved turbulent fluctuations must be modeled, and a particularly important manifestation of TCI is through their influence on the local mean chemical source term in RANS ($\langle S_{\alpha,\text{chem}} \rangle$), or on the local spatially filtered chemical source term in LES ($\langle S_{\alpha,\text{chem}} \rangle_\Delta$).

As an example, consider a single binary reaction involving two species (fuel F and oxidizer O) with zero temperature exponent [Eq. (2.3) through Eq. (2.5) with $b_{l,f} = 0$]. Then the chemical source term in the species mass fraction equation for product species P can be written as,

$$S_{P,\text{chem}} = A Y_F Y_O \exp\{-T_A/T\} \,, \tag{2.31}$$

where all necessary units conversions have been included in the pre-exponential factor A, and $T_A = E_A/R_U$ is the activation temperature for the reaction. The mean chemical source term in the RANS mean species equation is then,

$$\langle S_{P,\text{chem}} \rangle = A \langle Y_F Y_O \exp\{-T_A/T\} \rangle \,, \tag{2.32}$$

and clearly $\langle S_{P,\text{chem}} \rangle \neq A \langle Y_F \rangle \langle Y_O \rangle \exp\{-T_A/\langle T \rangle\}$ (the same functional form as in the unaveraged/unfiltered equation, expressed in terms of mean composition and mean temperature). The difference between $\langle S_{P,\text{chem}} \rangle$ and $A \langle Y_F \rangle \langle Y_O \rangle \exp\{-T_A/\langle T \rangle\}$ can be identified as the influence of turbulence–chemistry interactions on the mean chemical source term:

$$\begin{aligned} \langle S_{P,\text{chem}} \rangle = {} & A \langle Y_F \rangle \langle Y_O \rangle \exp\{-T_A/\langle T \rangle\} \\ & + \left[A \langle Y_F Y_O \exp\{-T_A/T\} \rangle - A \langle Y_F \rangle \langle Y_O \rangle \exp\{-T_A/\langle T \rangle\} \right] \,, \end{aligned} \tag{2.33}$$

where the term in square brackets then corresponds to TCI. We consider the constant-temperature case first, for simplicity. (Constant temperature is of little practical interest for combustion, but is relevant in some chemical engineering applications of turbulent reacting flows, for example.) In that case, the exponential term can be included in A, so that:

$$\begin{aligned} \langle S_{P,\text{chem}} \rangle &= A \langle Y_F \rangle \langle Y_O \rangle + \left[A \langle Y_F Y_O \rangle - A \langle Y_F \rangle \langle Y_O \rangle \right] \\ &= A \langle Y_F \rangle \langle Y_O \rangle + A \langle Y_F' Y_O' \rangle \,. \end{aligned} \tag{2.34}$$

That is, for the case of a binary constant-temperature reaction, TCI can be expressed in terms of the correlation between turbulent fluctuations in the reactant mass fractions, and a RANS-based model for TCI would essentially correspond to a model for this correlation. The situation is more complicated in the variable-temperature case. There one can introduce a Reynolds decomposition for the temperature ($T = \langle T \rangle + T'$) and for the species mass fractions in Eq. (2.33), expand the exponential term using a Taylor series, manipulate the result to express the term in square brackets as an infinite series involving correlations among fluctuations in Y_F, Y_O, and T, and introduce models for the various correlations that appear (see [15], for example). This direct-expansion-based approach to TCI modeling has been largely abandoned, however, as no satisfactory approach for truncating the infinite series or for modeling the individual correlations has been found. Additional difficulties arise in LES. Returning to the constant-temperature case again for illustrative purposes, the LES analogue to Eq. (2.34) for the spatially filtered chemical source term is:

$$\langle S_{P,\text{chem}} \rangle_\Delta = A \langle Y_F \rangle_\Delta \langle Y_O \rangle_\Delta + \left[A \langle Y_F Y_O \rangle_\Delta - A \langle Y_F \rangle_\Delta \langle Y_O \rangle_\Delta \right]$$
$$= A \langle Y_F \rangle_\Delta \langle Y_O \rangle_\Delta + A \langle Y'_F Y'_O \rangle_\Delta + A \langle \langle Y_F \rangle_\Delta Y'_O \rangle_\Delta$$
$$+ A \langle \langle Y_O \rangle_\Delta Y'_F \rangle_\Delta + A \big(\langle \langle Y_F \rangle_\Delta \langle Y_O \rangle_\Delta \rangle_\Delta - \langle Y_F \rangle_\Delta \langle Y_O \rangle_\Delta \big) \, . \quad (2.35)$$

That is, in addition to the correlation between subfilter-scale reactant-mass-fraction fluctuations, there are additional TCI contributions involving correlations between resolved-scale quantities and subfilter-scale fluctuations, and correlations between resolved-scale quantities. And again, the situation becomes more complicated with consideration of temperature fluctuations.

Local manifestations of TCI in turbulent flames can include local extinction/re-ignition of the flame. Global effects of TCI in both RANS and in LES for nonpremixed systems generally include reductions in the rate of heat release, lower peak mean temperatures, and significant ($>100\,\%$) changes in pollutant emissions. For sufficiently low values of the Damköhler number (ratio of a turbulence time scale to a chemical time scale), TCI can result in global extinction of the flame. That is, for conditions that correspond to global flame blowout in an experiment, a simulation that neglects TCI may result in a robustly burning flame, whereas a simulation that accounts properly for TCI would correctly predict global extinction.

A RANS example for a nonpremixed turbulent jet flame is provided in Fig. 2.2. The configuration is a round fuel jet of methane (diluted with nitrogen and other gases) issuing into ambient air, with an annular pilot to stabilize the flame; this is Sandia "Flame D" [17], which has been the subject of numerous experimental and modeling studies. In Fig. 2.2, computed mean temperature and mean O_2 mass fraction profiles computed using a composition PDF method are compared with profiles computed using a model that ignores turbulent fluctuations altogether (the local mean chemical source terms are computed based on the local mean species composition and temperature—a WSR model). Here a simple one-step, finite-rate global methane–air chemical mechanism has been used; good quantitative agreement with experiment therefore is not expected. Nevertheless, this serves to illustrate the benefits of accounting explicitly for turbulent fluctuations in species composition and temperature using a PDF method. The model that ignores turbulent fluctuations grossly overpredicts the local mean fuel- and oxygen-consumption rates (hence heat-release rate) and overpredicts local mean temperatures by as much as several hundred Kelvin. Results are significantly improved with the PDF model, in spite of the oversimplified chemistry.

There are analogous effects for the radiation source term, and turbulence–radiation interactions (TRI) will be introduced in Chap. 3. Further examples of TCI and TRI in RANS and LES modeling studies will be presented and discussed in subsequent chapters.

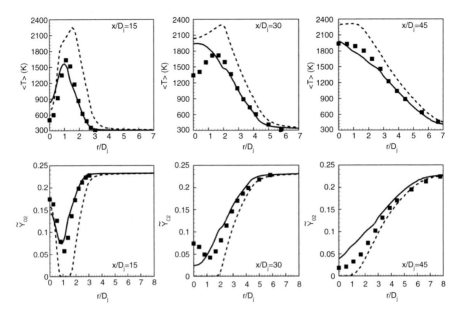

Fig. 2.2 Computed (*lines*) [16] and measured (*symbols*) [17] radial profiles of mean temperature (*upper row*) and mean O_2 mass fraction (*lower row*) at three axial locations in Sandia Flame D. Here D_j is the fuel-jet diameter. *Solid lines* are from a PDF-based model. *Dashed lines* are from a model that ignores the influence of turbulent fluctuations on mean chemical source terms (a WSR model). This is Fig. 8 of [8]

2.5 Multiphase Systems

Liquid and solid fuels are often used in practical combustion systems, and solid particulate matter (PM) is a key regulated pollutant. Therefore, radiative heat transfer in multiphase turbulent combustion is of interest. Discussions of the radiative properties of multiphase mixtures and how those are modeled are deferred to Chap. 3. Here we will introduce the basic physical processes that are considered in modeling soot particles, liquid sprays, and coal particles, and how those are implemented in the context of a transported composition PDF modeling framework. Soot (carbonaceous particles produced in the combustion process—a principal constituent of PM from combustion systems) is discussed first. Typical soot particle sizes are in the range 10–100 nm, and soot is typically modeled as a dilute continuum, where additional equations that are similar to the gas-phase species equations are solved to calculate the spatial and temporal distributions of soot quantities of interest (e.g., soot volume fraction). Liquid fuel sprays are discussed next. Typical fuel droplet diameters in combustion systems range from 1 to 100 μm. While continuum (Eulerian) model formulations can be used for liquid fuel injection and spray modeling, a stochastic Lagrangian parcel method (distinct from the stochastic Lagrangian particle method that is used to solve the modeled composition

PDF equation for the gas phase) usually is used. Finally, basic elements of coal combustion modeling are discussed. Coal particle sizes for practical combustion devices vary widely, and again, a separate stochastic Lagrangian parcel method (similar to the one that is used for spray modeling) is considered here.

2.5.1 Soot

In most cases, soot is an undesirable by-product of the combustion process. It has been receiving increasing scrutiny as an air pollutant, because of both its physical and its chemical properties. The adverse health effects of inhaling fine particles have been recognized, and toxic volatile compounds have been identified in soot. Soot provides a strong thermal signature in propulsion systems where detection is undesirable. Its role in global climate change is being debated. In industrial furnaces and glass melting, high levels of soot may be desirable to increase heat transfer rates by thermal radiation, but even there it is necessary to minimize the soot that is released to the environment. There are soot-based commercial materials, such as carbon blacks and some nanomaterials. Reviews of soot physics and modeling, and entry points to the literature, can be found in [18–24].

Until recently, particulate emissions from transportation vehicles and other applications have been regulated based on the total quantity (mass) of particles emitted, and the quantity of primary interest from a soot model has been the local soot mass fraction or volume fraction. Soot volume fraction is also the primary quantity of interest for calculating soot radiation (Chap. 3). For these reasons, the goal of most modeling studies has been to predict the distribution of soot volume fraction in a flame. More recently, there has been increasing concern about health effects of ultrafine particles, and new regulations based on the total number of particles emitted are being developed and phased in. This places increasing emphasis on predicting the particle size distribution function (PSDF). As a consequence of these new number-based particulate standards, soot modeling is of increasing interest in combustion systems where particulate emissions had not been a concern earlier: in particular, for premixed combustion systems including homogeneous-charge spark-ignition piston engines in automobiles.

The current understanding of soot formation in flames involves four basic steps (Fig. 2.3): inception of solid particles from high-molecular-weight gas-phase precursors (primarily polycyclic aromatic hydrocarbons—PAHs), mainly in locally fuel-rich regions of a flame; particle coagulation, where soot particles collide with one another and coalesce to form larger (essentially spherical) particles; heterogeneous surface reactions (growth and/or oxidation) with gas-phase species; and particle agglomeration, where soot particles collide with one another and coalesce to form chains of spherical particles (approximated as mass fractals). Typical primary (spherical) particle sizes for mature soot particles are on the order of 10 nm, and soot aggregates can grow to be as large as 1 μm. Acetylene is the principal gas-phase species that is involved in particle surface-growth reactions [26];

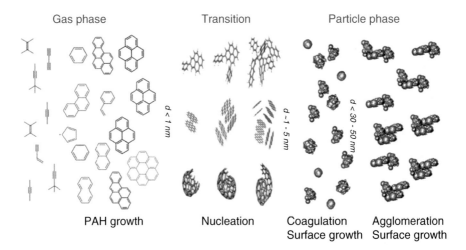

Fig. 2.3 A schematic illustration of soot formation processes in flames [25]

the surface-growth reactions are analogous to the gas-phase reactions that occur for large PAH molecules, and follow the H-abstraction-C_2H_2-addition (HACA) mechanism [27–29]. Surface condensation of high-molecular-weight gas molecules (e.g., PAHs) onto particles also has been considered [21]. Surface oxidation is principally via OH and O_2 [30], and occurs mainly in locally fuel-lean regions of a flame. Particle fragmentation (oxidation-induced fragmentation, where the oxidation of carbon in a soot particle results in breakup of the particle into smaller particles) also has been considered [31]. Simplifications that have been invoked to facilitate modeling include neglecting the internal physical structure and chemical composition of soot particles (e.g., assuming that they are homogeneous spheres of pure carbon), basing the particle inception rate on the local fuel or acetylene concentration (versus PAH concentration) [32], and neglecting agglomeration into nonspherical particles (generally satisfactory for atmospheric-pressure flames).

   This physical understanding has been incorporated to varying degrees into the models that have been applied to predict soot in laminar and turbulent flames [20]. The simplest models use empirical algebraic correlations to relate local soot volume fraction to local thermochemical conditions (e.g., equivalence ratio and temperature) [33, 34]. At the next level are semiempirical models. There one solves one or more PDEs for quantities such as particle number density and/or soot volume or mass fraction [32, 35], with simplified source/sink terms to account for the underlying physical subprocesses. Detailed models attempt to explicitly represent each of the key physical subprocesses described in the preceding paragraph. These generally require information about the distribution of soot particle sizes (the PSDF), and models for soot particle dynamics have been developed that are analogous to those that have been used for aerosol dynamics in other domains of application. Several approaches have been proposed for modeling soot particle dynamics. These include sectional methods where one solves modeled equations for discrete size classes (sections) [36], moment methods where one solves a truncated set of modeled equations for the lower-order moments of the PSDF [37], hybrids of sectional and moment methods [38], stochastic methods where Monte Carlo methods are used to solve for the PSDF [39], and PDF-based methods to solve for the PSDF [40].

Several PDF-based modeling studies of sooting turbulent flames have been reported using the Lagrangian-particle-based composition PDF methods described in Sect. 2.3 [41–45]. To capture the effects of turbulent fluctuations on soot processes, additional quantities are transported on each notional particle. For example, in a method-of-moments-based approach [41, 43, 44], the lower-order concentration moments of the soot PSDF are transported on each particle, and a closure model is invoked to truncate the infinite series of moments (e.g., interpolative closure [37]). Source terms in the moment equations are designed to represent each of the soot physical subprocesses described earlier. Only the lowest-order moments of the PSDF have simple physical interpretations: the zeroth-order moment is the particle number density, and the first-order moment is the particle mass per unit volume. Standard mixing models have been applied to soot moments in some PDF modeling studies [43, 44], although it has been argued that soot moments should not be mixed in the same manner as gas-phase species [41].

Examples of PDF-based modeling studies of luminous atmospheric-pressure laboratory-scale flames and of soot formation in diesel engines are discussed in Chaps. 6 and 7, respectively. There the importance of the highly nonlinear interactions among turbulence, gas-phase chemistry, soot, and thermal radiation is highlighted.

2.5.2 Liquid Fuel Sprays

In many turbulent combustion systems, liquid fuel is injected as a spray. Injected droplet sizes range from as small as a few microns for modern high-pressure diesel engines to $100\,\mu$m or more for low-pressure combustors. Reviews of the physics and modeling of sprays and spray combustion can be found in [46–48]. The fuel spray can be considered as a dispersed liquid phase in a carrier gas phase. Here the focus is on approaches that treat the carrier gas phase using the PDF methods that have been discussed earlier, and that have been applied to laboratory- and/or device-scale turbulent spray flames. Most of these modeling studies have used a separate stochastic Lagrangian representation for the dispersed liquid phase that follows the formulation introduced by Dukowicz [49]. Although alternative formulations are available for treating dispersed phases in turbulent reacting flows [50–52], the stochastic Lagrangian approach has been implemented in many research and commercial codes, and probably is the most widely used today for modeling chemically reacting turbulent flows with sprays. For these reasons, a stochastic Lagrangian spray formulation is taken as the baseline here.

Formally, the liquid phase is represented in terms of a droplet distribution function [53, 54] that is discretized into a number of computational parcels or particles, with each parcel representing a group of droplets that share the same properties. Some more recent models account for a distribution of droplet sizes within each computational parcel [55]. The number of computational spray parcels that can be used in a simulation generally is orders of magnitude smaller than the number of physical droplets in the system; a statistical representation of the droplet phase is introduced where a relatively small number of computational parcels is used to model the dynamics of the entire spray. The principal physical properties that are transported with each parcel usually include droplet velocity, droplet size (radius or diameter, for spherical droplets), droplet temperature, and the mass or number of physical droplets that the parcel represents. Additional quantities are required to represent multicomponent fuels that have different vaporization rates for different components [56]. Physical submodels that account for vaporization, droplet–droplet collisions, wall impingement, aerodynamic breakup, micro-explosions, etc., are implemented directly for each computational spray parcel (Fig. 2.4). Detailed

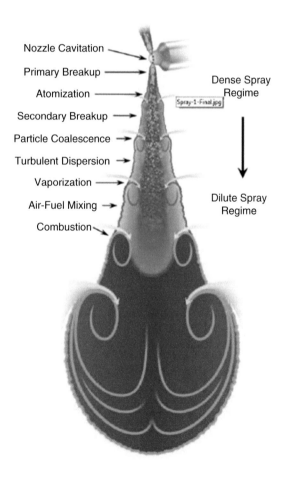

Fig. 2.4 A schematic illustration of liquid fuel injection and spray processes. Figure courtesy of Dr. Joseph Oefelein

Nozzle Cavitation

Primary Breakup

Atomization

Secondary Breakup

Particle Coalescence

Turbulent Dispersion

Vaporization

Air-Fuel Mixing

Combustion

Dense Spray Regime

Dilute Spray Regime

descriptions of state-of-the-art stochastic Lagrangian fuel spray models can be found in [57–59], for example, and a review of the theoretical foundations and modeling aspects of the approach can be found in [60].

It is important to recognize that the computational parcels that are used to represent the liquid fuel spray are not the same as the computational particles that are used to model and solve the gas-phase PDF equation. The reason is that the velocity of a spray droplet can be much different than the velocity of the local gas mixture. Thus a dual-Lagrangian formulation is used where one ensemble of particles represents the dispersed liquid phase (fuel parcels) and a separate ensemble of particles represents the carrier gas phase (notional PDF particles). With this representation, interactions between property fluctuations in the two phases can be captured in a natural way, and are not limited (in principle) to cell-level mean quantities. The principal liquid-to-gas-phase coupling is through the vaporization model. For example, new gas-phase particles having the correct mass, composition, and enthalpy (as determined by the vaporization model) can be introduced at the position of the vaporizing liquid parcel [61, 62]; this captures in a natural and direct manner the high level of local gas-phase concentration fluctuations that arise from vaporizing sprays, an effect that is either ignored or must be modeled in alternative (nonPDF) approaches [63, 64]. Alternatively, the mass and thermochemical properties of the vaporizing fuel can be distributed over nearby existing gas-phase particles [65]; comparisons of the two approaches are given in [66, 67]. These approaches are appropriate for an external group combustion regime where combustion occurs around groups of droplets (versus a flame around each individual droplet, for example). In most models only the mean gas-phase properties influence liquid-phase droplet behavior, although in principle the effects of gas-phase fluctuations could be applied more directly.

Composition-PDF modeling studies of turbulent reacting flows with liquid fuel sprays using dual-Lagrangian-particle formulations include [61, 62, 65–67]. Spray-radiation coupling is discussed in Sect. 7.3.

2.5.3 Coal

Coal combustion is a complex process, and the physical complexity is exacerbated by the large differences among coals of different types/ranks. This makes it difficult to extrapolate what is learned in one experiment or modeling study to different conditions where a different coal is used [68]. Key physical subprocesses in coal combustion include devolatilization, char combustion, heat transfer, and particle kinematics (Fig. 2.5).

Devolatilization is an endothermic process where heat transfer from the gas phase to coal particles increases the particle temperature, thereby causing the release of volatile gases including hydrocarbons and water vapor. Modeling studies [70, 71] show that computed temperatures and other global features are most sensitive to the devolatilization model that is used, because this process governs the amount

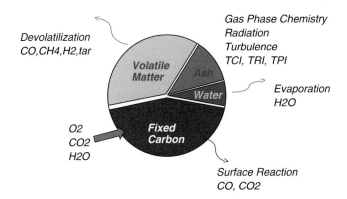

Fig. 2.5 A schematic of physical and chemical processes in coal combustion [69]

and rate of the release of fuel that will subsequently burn in the gas phase. The simplest devolatilization model is a constant-rate model [72], where volatile gases are released at a rate that is independent of temperature or composition. This model is sometimes used at the beginning of a simulation to initiate the flame [73], but does not yield satisfactory results for the subsequent combustion process. Devolatilization is usually modeled using Arrhenius-like kinetic expressions for the devolatilization rate. In single-rate kinetic models [74], the rate of devolatilization has a first-order dependence on the concentration of volatiles in the coal particle and an exponential (Arrhenius) dependence on temperature. Model parameters are derived from experiments or from higher-order devolatilization models, and depend both on the specific coal type and on the combustion environment (e.g., the local heating rate). Two-rate models include two parallel reaction pathways for low- and high-temperature ranges [75], and are formulated and calibrated similarly to single-rate models. Single- and two-rate models are computationally efficient, but are limited to a narrow range of conditions for which they have been calibrated. The distributed-activation-energy model (DAEM) [76] uses multiple parallel first-order reactions, and an average activation energy with a spread of energies described by a Gaussian distribution is used, with a specified standard deviation. All of these kinetics-based models implicitly include the notion of breaking chemical bonds that have a specified activation energy.

In contrast, network models correlate the devolatilization rate with the chemical structures and functional groups [73], and they do not depend on first-order reaction kinetics. These models require detailed input concerning kinetic parameters, composition, and polymeric structure of the coal of interest. Some of the network models provide different devolatilization rates for individual species, and they can be computationally intensive. Three major network models are available: the FLASHCHAIN model [77–79], the Chemical Percolation Devolatilization (CPD) model [80–82], and the Functional Group-Depolymerization Vaporization Cross-linking (FG-DVC) model [83]. These models are usually implemented in a pre-processing step with the properties of the coal of interest as the input.

The Carbonaceous Chemistry for Computational Modeling (C3M) software from the National Energy Technology Laboratory allows the devolatilization rate parameters calculated from a network model to be integrated into a CFD model.

In addition to the devolatilization rate(s), the composition of the volatile gases (light gases and tar) also must be specified in a devolatilization model. It is desirable to predict the molecular weight distribution of the devolatilized gases, and such information is available from network models. A common practice in CFD is to adopt a more expedient approach of a "pseudomolecule" having an appropriate molecular weight [84], based on the proximate (percentages of individual species) and ultimate (element percentages) analysis of the coal. To simplify the gas-phase chemistry that needs to be considered, it is often further assumed that the volatile gases break down instantaneously to form low-molecular-weight species (e.g., CH_4, CO, C_2H_4, CO_2, H_2O).

Heterogeneous char reactions are slow compared to devolatilization. In pulverized coal/air combustion, a typical time scale for devolatilization is ~ 0.1 s, while a typical char burnout time is ~ 1 s [85]. The underlying physical processes include pore diffusion, adsorption, complex formation, rearrangement, and desorption [85], and the overall reaction rate is determined by the particle temperature, local partial pressure of oxygen, particle size, and porosity. Most kinetic data for char combustion have been expressed using a single-film model with an nth-order Arrhenius form [86]. Depending on which step is rate-limiting (slowest), three regimes can be identified based on the local temperature: kinetic controlled, diffusion-kinetic controlled, and diffusion controlled [3]. For example, in the diffusion-controlled regime, surface reactions proceed at a rate that is determined by the rate of diffusion of the gases to the surface of the char, and the kinetics of the surface reaction are assumed to be fast. It has been shown that the combustion of most particles in pulverized coal/air flames occurs in the diffusion-kinetic controlled regime [73]. The available models can be categorized as either global models or intrinsic models. Global models do not account explicitly for char porosity, internal diffusion of oxygen and internal reactions; that is, those effects are lumped into an effective kinetic rate of char oxidation. In contrast, in intrinsic models the porosities and internal reactions are separated from the intrinsic kinetic reaction rate. Global models are designed for a single regime, and the global reaction $C + O_2 \rightarrow CO_2$ or $C + 0.5\ O_2 \rightarrow CO$ is considered as the sole heterogeneous reaction. Further simplifications include assumptions of uniform properties and spherical particles. Global models are simple and straightforward to implement, and are used in most cases where the temperature is higher than 1500 K. The intrinsic model proposed by Smith [85] is an example of an advanced char combustion model, where rate parameters explicitly consider the internal pore diffusion and internal reactions. The carbon burnout kinetics (CBK) model [87] is a variant of an intrinsic model that was designed to predict the total extent of carbon burnout and ultimate fly ash carbon content for prescribed temperature/oxygen history. Other variants of carbon burnout kinetics models consider gasification reactions.

When the particle phase is sufficiently dispersed (volume fraction less than 10 %, so that particle–particle interactions are negligible), a stochastic Lagrangian parcel method is usually adopted for the particle phase, similar to the one described earlier for liquid fuel sprays. As was the case for sprays, a sampling method is used where the number of computational coal parcels is much smaller than the number of physical coal particles, and each parcel represents a large number of coal particles that have the same properties. Physical processes that are considered include drag forces, buoyancy forces, particle dispersion due to turbulent velocity fluctuations, and heat and mass transfer between particle and gas phases (based on the devolatilization and char reaction models described earlier). Common simplifications include spherical particle assumptions, and neglecting effects such as particle swelling that can occur during the devolatilization process, thermophoresis, and particle–particle interactions. The coupling between the stochastic parcel method that is used to represent the particle phase and the stochastic particle method that is used to represent the gas phase in the context of a transported composition PDF method is similar to that described earlier for liquid fuel sprays. In the case of coal, the coupling is more complicated, as multiple species are involved (in contrast to a single species in the case of sprays—fuel) and there is two-way mass transfer between phases because of the heterogeneous reactions (in contrast to the one-way mass transfer corresponding to vaporization in the case of sprays).

Finally, it is noted that most work to date has been in the context of pulverized coal–air combustion systems. There is growing interest in oxy-coal combustion, where pure oxygen is used as an oxidizer instead of air, to facilitate CO_2 capture and sequestration [70, 73, 86]. This requires modifications to some of the physical submodels. Of particular interest here is high-temperature oxy-coal combustion, where flue-gas recirculation is not used, so that peak temperatures approach 3000 K, thereby significantly altering the heat-transfer environment compared to conventional air–coal combustion or oxy-coal with flue-gas recirculation where peak temperatures are $\sim$2100 K. Examples of pulverized coal and oxy-fuel combustion are discussed in Sect. 6.5.

References

1. K.K. Kuo, *Principles of Combustion*, 2nd edn. (Wiley, Hoboken, 2005)
2. R.J. Kee, G. Dixon-Lewis, J. Warnatx, M.E. Coltrin, J.A. Miller, A Fortran computer code package for the evaluation of gas-phase, multicomponent transport properties. Technical Report SAND86-8246, Sandia National Laboratory, 1986
3. S.R. Turns, *An Introduction to Combustion: Concepts and Applications*, 3rd edn. (McGraw-Hill, New York, 2011)
4. R.S. Barlow, J.H. Frank, Effects of turbulence on species mass fractions in methane/air jet flames. Proc. Combust. Inst. **27**, 1087–1095 (1998)
5. D. Veynante, J. Piana, J.M. Duclos, C. Martel, Experimental analysis of flame surface density models for premixed turbulent combustion. Proc. Combust. Inst. **26**, 413–420 (1996)
6. T.D. Fansler, D.T. French, The scavenging flow field in a crankcase-compression two-stroke engine: a three-dimensional laser-velocimetry survey. SAE Technical Paper No. 920417 (1992)
7. S.B. Pope, PDF methods for turbulent reactive flows. Prog. Energy Combust. Sci. **11**, 119–192 (1985)

8. D.C. Haworth, Progress in probability density function methods for turbulent reacting flows. Prog. Energy Combust. Sci. **36**, 168–259 (2010)
9. D.C. Haworth, S.B. Pope, Transported probability density function methods for Reynolds-averaged and large-eddy simulations, in *Turbulent Combustion Modeling - Advances, New Trends and Perspectives*, ed. by T. Echekki, E. Mastorakos (Springer, Berlin, 2011), pp. 119–142
10. S.B. Pope, Computations of turbulent combustion: progress and challenges. Proc. Combust. Inst. **23**, 591–612 (1990)
11. P. Givi, Filtered density function for subgrid scale modeling of turbulent combustion. Am. Inst. Aeronaut. Astronaut. J. **44**, 16–23 (2006)
12. A. Leonard, Energy cascade in large eddy simulation of turbulent fluid flow. Adv. Geophys. **18A**, 237–248 (1974)
13. S.B. Pope, *Turbulent Flows* (Cambridge University Press, Cambridge, 2000)
14. J. Jaishree, D.C. Haworth, Comparisons of Lagrangian and Eulerian PDF methods in simulations of nonpremixed turbulent jet flames with moderate-to-strong turbulence-chemistry interactions. Combust. Theory Model. **16**, 435–463 (2012)
15. T. Poinsot, D. Veynante, *Theoretical and Numerical Combustion*, 3rd edn. (Poinsot, Toulouse, 2011)
16. Y.Z. Zhang, Hybrid particle/finite-volume PDF methods for three-dimensional time-dependent flows in complex geometries. Ph.D. thesis, The Pennsylvania State University, University Park, 2004
17. R.S. Barlow, J.H. Frank, Effects of turbulence on species mass fractions in methane/air jet flames. Proc. Combust. Inst. **27**, 1087–1095 (1998)
18. B.S. Haynes, H.G. Wagner, Soot formation. Prog. Energy Combust. Sci. **7**, 229–273 (1981)
19. H. Bockhorn (ed.), *Soot Formation in Combustion: Mechanisms and Models* (Springer, Berlin, 1994)
20. I.M. Kennedy, Models of soot formation and oxidation. Prog. Energy Combust. Sci. **23**, 95–132 (1997)
21. H. Richter, J.B. Howard, Formation of polycyclic aromatic hydrocarbons and their growth to soot - a review of chemical reaction pathways. Prog. Energy Combust. Sci. **26**, 565–608 (2000)
22. D.R. Tree, K.I. Svensson, Soot processes in compression ignition engines. Prog. Energy Combust. Sci. **33**, 272–309 (2007)
23. C.R. Shaddix, T.C. Williams, Soot: giver and taker of light. Am. Sci. **95**, 232–239 (2007)
24. H. Wang, Formation of nascent soot and other condensed-phase materials in flames. Proc. Combust. Inst. **33**, 41–67 (2011)
25. S.P. Roy, Aerosol-dynamics-based soot modeling of flames. Ph.D. thesis, The Pennsylvania State University, University Park, 2014
26. S.J. Harris, A.M. Weiner, Chemical kinetics of soot particle growth. Ann. Rev. Phys. Chem. **36**, 31–52 (1985)
27. M. Frenklach, H. Wang, Detailed modeling of soot particle nucleation and growth. Proc. Combust. Inst. **23**, 1559–1566 (1991)
28. M. Frenklach, H. Wang, Detailed mechanism and modeling of soot particle formation, in *Soot Formation in Combustion: Mechanisms and Models*, ed. by H. Bockhorn (Springer, Berlin, 1994), pp. 162–190
29. M. Frenklach, Reaction mechanism of soot formation in flames. Phys. Chem. Chem. Phys. **4**, 2028–2037 (2002)
30. K.G. Neoh, J.B. Howard, A.F. Sarofim, Soot oxidation in flames, in *Particulate Carbon: Formation During Combustion*, ed. by D.C. Siegla, G.W. Smith (Plenum, New York, 1981), pp. 162–282
31. M. Sirignano, J. Kent, A. D'Anna, Modeling formation and oxidation of soot in nonpremixed flames. Energy Fuels **27**, 2303–2315 (2013)
32. K.M. Leung, R.P. Lindstedt, W.P. Jones, A simplified reaction mechanism for soot formation in nonpremixed flames. Combust. Flame **87**, 289–305 (1991)

33. J.P. Gore, G.M. Faeth, Structure and spectral radiation properties of turbulent ethylene/air diffusion flames. Proc. Combust. Inst. **21**, 1521–1531 (1986)
34. S.Y. Lee, Detailed studies of spatial soot formation processes in turbulent ethylene jet flames. Ph.D. thesis, The Pennsylvania State University, University Park, 1998
35. I.M. Kennedy, W. Kollmann, J.Y. Chen, A model for soot formation in a laminar diffusion flame. Combust. Flame **81**, 73–85 (1990)
36. M.B. Colket, R.J. Hall, Successes and uncertainties in modelling soot formation in laminar, premixed flames, in *Soot Formation in Combustion: Mechanisms and Models* (Springer, Berlin, 1994), pp. 442–468
37. M. Frenklach, Method of moments with interpolative closure. Chem. Eng. Sci. **57**, 2229–2239 (2002)
38. M.E. Mueller, G. Blanquart, H. Pitsch, Hybrid method of moments for modeling soot formation and growth. Combust. Flame **156**, 1143–1155 (2009)
39. M. Balthasar, M. Kraft, A stochastic approach to calculate the particle size distribution function of soot particles in laminar premixed flames. Combust. Flame **133**, 289–298 (2003)
40. S. Rigopoulos, PDF method for population balance in turbulent reactive flow. Chem. Eng. Sci. **62**, 6865–6878 (2007)
41. R.S. Mehta, Detailed modeling of soot formation and turbulence-radiation interactions in turbulent jet flames. Ph.D. thesis, The Pennsylvania State University, University Park, 2008
42. W. Kollmann, I.M. Kennedy, M. Metternich, J.-Y. Chen, PDF prediction of sooting turbulent flames, in *Soot Formation in Combustion: Mechanisms and Models*, ed. by H. Bockhorn (Springer, Berlin, 1994), pp. 503–526
43. M. Balthasar, F. Mauss, A. Knobel, M. Kraft, Detailed modeling of soot formation in a partially stirred plug flow reactor. Combust. Flame **128**, 395–409 (2002)
44. R.P. Lindstedt, S.A. Louloudi, Joint-scalar transported PDF modeling of soot formation and oxidation. Proc. Combust. Inst. **30**, 775–783 (2005)
45. I.M. Aksit, J.B. Moss, A hybrid scalar model for sooting turbulent flames. Combust. Flame **145**, 231–244 (2006)
46. G.M. Faeth, Spray combustion phenomena. Proc. Combust. Inst. **26**, 1593–1612 (1996)
47. W.A. Sirignano, *Fluid Dynamics and Transport of Droplets and Sprays* (Cambridge University Press, Cambridge, 1999)
48. P. Jenny, D. Roekaerts, N. Beishuizen, Modeling of turbulent dilute spray combustion. Prog. Energy Combust. Sci. **38**, 846–887 (2012)
49. J.K. Dukowicz, A particle-fluid numerical model for liquid sprays. J. Comp. Phys. **35**, 229–253 (1980)
50. E. Loth, Numerical approaches for motion of dispersed particles, droplets and bubbles. Prog. Energy Combust. Sci. **26**, 161–223 (2000)
51. F. Mashayek, R.V.R. Pandya, Analytical description of particle/droplet-laden turbulent flows. Prog. Energy Combust. Sci. **29**, 329–378 (2003)
52. D.L. Marchisio, R.O. Fox, *Multiphase Reacting Flows: Modelling and Simulation* (Springer, Berlin, 2007)
53. F.A. Williams, Spray combustion and atomization. Phys. Fluids **1**, 541–545 (1958)
54. S. Subramaniam, Statistical modeling of sprays using the droplet distribution function. Phys. Fluids **13**, 624–642 (2001)
55. A. Kösters, A. Karlsson, Validation of the VSB2 spray model against spray A and spray H. Atomization Sprays (2015, in press)
56. A.M. Lippert, R.D. Reitz, Modeling of multicomponent fuels using continuous distributions with application to droplet evaporation and sprays. SAE Technical Paper No. 972882 (1997)
57. M.S. Raju, LSPRAY-II: a Lagrangian spray module. Technical Report NASA/CR-2004-212958. NASA Glenn Research Center, 2004
58. M.S. Raju, Numerical investigation of various atomization models in the modeling of a spray flame. Technical Report NASA/CR-2005-214033. NASA Glenn Research Center, 2005
59. C.M. Cha, J. Zhu, K. Rizk, M.S. Anand, A comprehensive liquid fuel injection model for CFD simulations of gas turbine combustors. AIAA paper no. 2005-0349 (2005)

60. S. Subramaniam, Lagrangian–Eulerian methods for multiphase flows. Prog. Energy Combust. Sci. **39**, 215–245 (2013)
61. Y.Z. Zhang, E.H. Kung, D.C. Haworth, A PDF method for multidimensional modeling of HCCI engine combustion: effects of turbulence/chemistry interactions on ignition timing and emissions. Proc. Combust. Inst. **30**, 2763–2771 (2005)
62. S. James, M.S. Anand, S.B. Pope, The Lagrangian PDF transport method for simulations of gas turbine combustor flows. AIAA Paper no. 2002-4017 (2002)
63. J. Réveillon, L. Vervisch, Spray vaporization in nonpremixed turbulent combustion modeling: a single droplet model. Combust. Flame **121**, 75–90 (2000)
64. J. Réveillon, L. Vervisch, Analysis of weakly turbulent dilute-spray flames and spray combustion regimes. J. Fluid Mech. **537**, 317–347 (2005)
65. H.-W. Ge, E. Gutheil, Simulation of a turbulent spray flame using coupled PDF gas phase and spray flamelet modeling. Combust. Flame **153**, 173–185 (2008)
66. E.H. Kung, D.C. Haworth, Transported probability density function (tPDF) modeling for direct-injection internal combustion engines. SAE Int. J. Engines **1**, 591–606 (2008)
67. E.H. Kung, PDF-based modeling of autoignition and emissions for advanced direct-injection engines. Ph.D. thesis, The Pennsylvania State University, University Park, 2008
68. L. Chen, S.Z. Yong, A.F. Ghoniem, Oxy-fuel combustion of pulverized coal: characterization, fundamentals, stabilization and CFD modeling. Prog. Energy Combust. Sci. **38**, 156–214 (2012)
69. X. Zhao, Transported probability density function methods for coal combustion: toward high temperature Oxy-coal for direct power extraction. Ph.D. thesis, The Pennsylvania State University, University Park, 2014
70. P. Nakod, G. Krishnamoorthy, M. Sami and S. Orsino, A comparative evaluation of gray and non-gray radiation modeling strategies in oxy-coal combustion simulations. Appl. Therm. Eng. **54**, 422–432 (2013)
71. J.D. Smith, P.J. Smith, S.C. Hill, Parametric sensitivity study of a CFD-based coal combustion model. Am. Inst. Chem. Eng. J. **39**, 1668–1679 (1993)
72. M.M. Baum, P.J. Street, Predicting the combustion behavior of coal particles. Combust. Sci. Technol. **3**, 231–243 (1971)
73. P. Edge, M. Gharebaghi, R. Irons, R. Porter, R.T.J. Porter, M. Pourkashanian, D. Smith, P. Stephenson, A. Williams, Combustion modelling opportunities and challenges for oxy-coal carbon capture technology. Chem. Eng. Res. Des. **89**, 1470–1493 (2011)
74. S. Badzioch, P.G.W. Hawksley, Kinetics of thermal decomposition of pulverized coal particles. Ing. Eng. Chem. Process Des. Dev. **9**, 521–530 (1970)
75. H. Kobayashi, J.B. Howard, A.F. Sarofim, Coal devolatilization at high temperatures. Proc. Combust. Inst. **16**, 411–425 (1977)
76. D. Anthony, J. Howard, H. Hottel, H. Meissner, Rapid devolatilization of pulverized coal. Proc. Combust. Inst. **15**, 1303–1317 (1975)
77. S. Niksa, FLASHCHAIN theory for rapid coal devolatilization kinetics. 1. Formulation. Energy Fuels **5**, 647–665 (1991)
78. S. Niksa, FLASHCHAIN theory for rapid coal devolatilization kinetics. 2. Impact of operating conditions. Energy Fuels **5**, 665–673 (1991)
79. S. Niksa, FLASHCHAIN theory for rapid coal devolatilization kinetics. 3. Modeling the behavior of various coals. Energy Fuels **5**, 673–683 (1991)
80. D.M. Grant, R.J. Pugmire, T.H. Fletcher, A.R. Kerstein, Chemical model of coal devolatilization using percolation lattice statistics. Energy Fuels **3**, 175–186 (1989)
81. T.H. Fletcher, A.R. Kerstein, R.J. Pugmire, D.M. Grant, Chemical percolation model for devolatilization: 2. Temperature and heating rate effects on product yields. Energy Fuels **4**, 54–60 (1990)
82. T.H. Fletcher, A.R. Kerstein, R.J. Pugmire, M.S. Solum, D.M. Grant, Chemical percolation model for devolatilization. 3. Direct use of 13C NMR data to predict effects of coal type. Energy Fuels **6**, 414–431 (1992)

83. P.R. Solomon, D.G. Hamblen, R.M. Carangelo, M.A. Serio, G.V. Deshpande, General model of coal devolatilization. Energy Fuels **2**, 405–422 (1988)
84. I. Petersen, J. Werther, Experimental investigation and modeling of gasification of sewage sludge in the circulating fluidized bed. Chem. Eng. Proc. **44**, 717–736 (2005)
85. I.W. Smith, The combustion rates of coal chars: a review. Proc. Combust. Inst. **19**, 1045–1065 (1982)
86. J.J. Murphy, C.R. Shaddix, Combustion kinetics of coal chars in oxygen-enriched environment. Combust. Flame **144**, 710–729 (2006)
87. R. Hurt, J.-K. Sun, M. Lunden, A kinetic model of carbon burnout in pulverized coal combustion. Combust. Flame **113**, 181–197 (1998)

Chapter 3
Radiation Properties, RTE Solvers, and TRI Models

3.1 Fundamentals of Thermal Radiation

Radiative heat transfer or *thermal radiation* is the science of transferring energy in the form of electromagnetic waves. Unlike heat conduction, electromagnetic waves do not require a medium for their propagation. Therefore, because of their ability to travel across vacuum, thermal radiation becomes the dominant mode of heat transfer in low pressure (vacuum) and outer space applications. Another distinguishing characteristic between conduction (and convection, if aided by flow) and thermal radiation is their temperature dependence. While conductive and convective fluxes are more or less linearly dependent on temperature differences, radiative heat fluxes tend to be proportional to differences in the fourth power of temperature (or even higher). For this reason, radiation tends to become the dominant mode of heat transfer in high-temperature applications, such as combustion (fires, furnaces, rocket nozzles), nuclear reactions (solar emission, nuclear weapons), and others.

All materials continuously emit and absorb electromagnetic waves, or photons, by changing their internal energy on a molecular level. Strength of emission and absorption of radiative energy depends on the temperature of the material, as well as on the wavelength λ, frequency v, or wavenumber η, that characterizes the electromagnetic waves:

$$\lambda = c/v = 1/\eta, \tag{3.1}$$

where wavelength is usually measured in μm ($= 10^{-6}$ m), while frequency is measured in Hz ($=$ cycles/s), and wavenumbers are given in cm^{-1}. Electromagnetic waves or photons (which include what is perceived as "light") travel at the *speed of light, c*. The speed of light depends on the medium through which the wave travels, and is related to that in vacuum, c_0, through the relation

© The Author(s) 2016
M.F. Modest, D.C. Haworth, *Radiative Heat Transfer in Turbulent Combustion Systems*, SpringerBriefs in Applied Sciences and Technology,
DOI 10.1007/978-3-319-27291-7_3

$$c = \frac{c_0}{n}, \qquad c_0 = 2.998 \times 10^8 \, \text{m/s}, \tag{3.2}$$

where n is known as the *refractive index* of the medium. By definition, the refractive index of vacuum is $n \equiv 1$. For most gases the refractive index is very close to unity, and the c in Eq. (3.1) can be replaced by c_0. Each wave or photon carries with it an amount of energy, ϵ, determined from quantum mechanics as

$$\epsilon = h\nu, \qquad h = 6.626 \times 10^{-34} \, \text{J-s}, \tag{3.3}$$

where h is known as *Planck's constant*. The frequency of light does not change when light penetrates from one medium to another because the energy of the photon must be conserved. On the other hand, the wavelength does change, depending on the values of the refractive index for the two media.

When an electromagnetic wave strikes an interface between two media, the wave is either reflected or transmitted. Most solid and liquid media absorb all incoming radiation over a very thin surface layer. Such materials are called *opaque*, or "opaque surfaces" (even though absorption takes place over a thin layer). An opaque material, that does not reflect any radiation at its surface, is called a "perfect absorber," or a "black surface" or a "blackbody," because such a surface appears "black" to the human eye, which recognizes objects by visible radiation reflected off their surfaces.

Emissive Power The rate with which a medium emits electromagnetic radiation into all directions depends on the local temperature and on the properties of the material. The radiative heat flux emitted from a surface is called the *emissive power*, E, and there is a distinction between *total* and *spectral emissive power* (heat flux emitted over the entire spectrum, or at a given wavelength per unit wavelength interval). Spectral and total emissive powers are related by

$$E(T) = \int_0^\infty E_\lambda(T, \lambda) \, d\lambda. \tag{3.4}$$

A black surface is not only a perfect absorber, but also a perfect emitter, that is, the emission from such a surface exceeds that of any other surface at the same temperature (known as *Kirchhoff's Law*). The emissive power leaving an opaque black surface, commonly called "blackbody emissive power," can be determined from quantum statistics as

$$E_{b\lambda}(T, \lambda) = \frac{2\pi hc_0^2}{n^2 \lambda^5 \left[e^{hc_0/n\lambda kT} - 1 \right]}, \qquad (n = \text{const}), \tag{3.5}$$

where it is assumed the black surface is adjacent to a nonabsorbing medium of constant refractive index n. The constant $k = 1.3806 \times 10^{-23}$ J/K is known as *Boltzmann's constant*. The spectral dependence of the blackbody emissive power

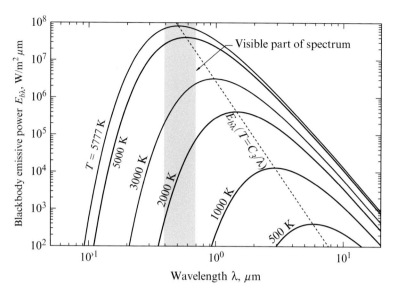

Fig. 3.1 Blackbody emissive power spectrum [1]

into vacuum ($n = 1$) is shown for a number of emitter temperatures in Fig. 3.1. It is seen that emission is zero at both extreme ends of the spectrum with a maximum at some intermediate wavelength. The general level of emission rises with temperature, and the important part of the spectrum (the part containing most of the emitted energy) shifts toward shorter wavelengths. Because emission from the sun ("solar spectrum") is well approximated by blackbody emission at an *effective solar temperature* of $T_{sun} = 5777$ K, this temperature level is also included in the figure. Combustion problems generally involve temperature levels between 500 K and, say, 2500 K. Therefore, the spectral ranges of interest in combustion applications are the near- and mid-infrared (0.7–20 μm).

Equation (3.5) has its maximum at

$$(n\lambda T)_{max} = 2898 \, \mu\text{m-K} \tag{3.6}$$

which is known as *Wien's displacement law*.

The total blackbody emissive power is found by integrating Eq. (3.5) over the entire spectrum, resulting in

$$E_b(T) = n^2 \sigma T^4, \tag{3.7}$$

where $\sigma = 5.670 \times 10^{-8}$ W/m²-K⁴ is the *Stefan–Boltzmann constant*.

Solid Angles Radiation is a directional phenomenon, that is, the radiative flux passing through a point generally varies with direction, such as the sun shining onto Earth from essentially a single direction. This is best understood by considering an

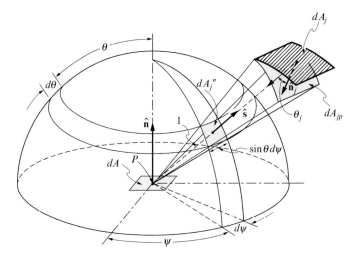

Fig. 3.2 Definitions of direction vectors and solid angles [1]

opaque (real or imaginary) surface element dA_i, as shown in Fig. 3.2. It is customary
to describe the direction unit vector $\hat{s}$ in terms of polar angle θ (measured from the
surface normal $\hat{n}$) and azimuthal angle ψ (measured in the plane of the surface,
between an arbitrary axis and the projection of $\hat{s}$); for a hemisphere $0 \leq \theta \leq \pi/2$
and $0 \leq \psi \leq 2\pi$.

The solid angle with which a surface A_j is seen from a certain point P (or dA_i
in Fig. 3.2) is defined as the projection of the surface onto a plane normal to the
direction vector, divided by the distance squared, as also shown in Fig. 3.2 for an
infinitesimal element dA_j. If the surface is projected onto a unit sphere above point
P, the solid angle becomes equal to the projected area, or

$$\Omega = \int_{A_{jp}} \frac{dA_{jp}}{S^2} = \int_{A_j} \frac{\cos \theta_0 dA_j}{S^2} = A''_{jp}, \quad (3.8)$$

where S is the distance between P and dA_j. Thus, an infinitesimal solid angle is
simply an infinitesimal area on a unit sphere, or

$$d\Omega = dA''_{jp} = (1 \times \sin \theta d\psi)(1 \times d\theta) = \sin \theta d\theta d\psi. \quad (3.9)$$

Because of its visual appearance $d\Omega$ is commonly referred to as a "pencil of rays."
Integrating over all possible directions yields

$$\int_{\psi=0}^{2\pi} \int_{\theta=0}^{\pi/2} \sin \theta d\theta d\psi = 2\pi \quad (3.10)$$

for the total solid angle above the surface. If a point inside a medium removed from the surface is considered, radiation passing through that point can strike any point of an imaginary unit sphere surrounding it, that is, the total solid angle here is 4π, with $0 \le \theta \le \pi, 0 \le \psi \le 2\pi$.

Radiative Intensity The directional behavior of radiative energy traveling through a medium is characterized by the radiative intensity I, which is defined as

$$I \equiv \text{radiative energy flow/time/area normal to the rays/solid angle.}$$

Like emissive power, intensity is defined on both spectral and total bases, related by

$$I(\hat{s}) = \int_0^\infty I_\lambda(\hat{s}, \lambda) \, d\lambda. \tag{3.11}$$

However, unlike emissive power, which depends only on position (and wavelength), the radiative intensity depends, in addition, on the direction vector $\hat{s}$. Emissive power can be related to emitted intensity by integrating this intensity over the 2π solid angles above a surface, and realizing that the projection of dA normal to the rays is $dA \cos\theta$. Thus,

$$E = \int_0^{2\pi} \int_0^{\pi/2} I(\theta, \psi) \cos\theta \, \sin\theta \, d\theta \, d\psi = \int_{2\pi} I(\hat{s}) \, \hat{n} \cdot \hat{s} \, d\Omega \tag{3.12}$$

which is, of course, also valid on a spectral basis. For a black surface it is readily shown that $I_{b\lambda}$ is independent of direction, or

$$I_{b\lambda} = I_{b\lambda}(T, \lambda). \tag{3.13}$$

Using this relation in Eq. (3.12), it is observed that the intensity leaving a blackbody (or any surface whose outgoing intensity is independent of direction, or *diffuse*) may be evaluated from the blackbody emissive power (or outgoing heat flux) as

$$I_{b\lambda} = E_{b\lambda}/\pi. \tag{3.14}$$

In the literature the spectral blackbody intensity is sometimes referred to as the *Planck function*. Note that intensity remains unchanged as it travels through vacuum (but the solid angle with which an emitting body is seen diminishes with increasing distance).

Radiative Heat Flux Emissive power is the total radiative energy streaming away from a surface due to emission. Therefore, it is a radiative flux, but not the net radiative flux at the surface, because it only accounts for emission and not for incoming radiation and reflected radiation. Extending the definition of Eq. (3.12), gives

$$(q_\lambda)_{\text{out}} = \int_{\cos\theta>0} I_\lambda(\theta)\cos\theta\, d\Omega \geq 0, \tag{3.15}$$

where $I_\lambda(\theta)$ is now *outgoing* intensity (due to emission plus reflection). Similarly, for incoming directions ($\pi/2 < \theta \leq \pi$)

$$(q_\lambda)_{\text{in}} = \int_{\cos\theta<0} I_\lambda(\theta)\cos\theta\, d\Omega < 0. \tag{3.16}$$

Combining the incoming and outgoing contributions, the net radiative flux at a surface is

$$(q_\lambda)_{\text{net}} = \mathbf{q}_\lambda \cdot \hat{\mathbf{n}} = (q_\lambda)_{\text{in}} + (q_\lambda)_{\text{out}} = \int_{4\pi} I_\lambda(\hat{\mathbf{s}})\cos\theta\, d\Omega. \tag{3.17}$$

The total radiative flux, finally, is obtained by integrating Eq. (3.17) over the entire spectrum, or

$$\mathbf{q} \cdot \hat{\mathbf{n}} = \int_0^\infty \mathbf{q}_\lambda \cdot \hat{\mathbf{n}}\, d\lambda = \int_0^\infty \int_{4\pi} I_\lambda(\hat{\mathbf{s}})\, \hat{\mathbf{n}} \cdot \hat{\mathbf{s}}\, d\Omega\, d\lambda. \tag{3.18}$$

Of course, the surface described by the unit vector $\hat{\mathbf{n}}$ may be an imaginary one (located somewhere inside a radiating medium). Thus, removing the $\hat{\mathbf{n}}$ from Eq. (3.18) gives the definition of the radiative heat flux vector inside a participating medium:

$$\mathbf{q} = \int_0^\infty \mathbf{q}_\lambda\, d\lambda = \int_0^\infty \int_{4\pi} I_\lambda(\hat{\mathbf{s}})\, \hat{\mathbf{s}}\, d\Omega\, d\lambda. \tag{3.19}$$

It is the divergence of the radiative flux that enters the overall energy equation in the form of a "radiative heat source" as given in Eq. (1.3)

$$S_{\text{rad}} = -\nabla \cdot \mathbf{q}_{\text{rad}}. \tag{3.20}$$

3.2 Radiative Properties of Combustion Systems

When a gas molecule absorbs or emits radiative energy, this raises or lowers the vibrational and/or rotational energies of the molecule. Since these energy levels are quantized this leads to many thousands of discrete spectral lines, forming so-called vibration–rotation bands in the infrared. The precise photon energies required for these transitions are altered a little by a number of effects, primarily due to molecular collisions and molecular movement (Doppler effect), leading to slight broadening of the spectral lines. A single spectral line at a certain spectral position is fully characterized by its strength and its line width (plus knowledge of the broadening mechanism, i.e., collision and/or Doppler broadening). Locations, strengths, and widths of spectral lines have been collected in modern databases, notably the HITRAN, HITEMP [2, 3], and CDSD [4] (for CO_2) databases, which also contain directions on how to calculate the resulting absorption coefficient. An example was already given in Fig. 1.2, showing the pressure-based absorption coefficient $\kappa_{p\eta}$ for the most important wavenumber range of the strong 4.3 μm band of carbon dioxide. The strong spectral variations of the absorption coefficient, covering multiple orders of magnitude, are clearly visible in the form of about 50 dominant broadened lines, although the given range contains more than 5000 lines in the HITRAN database (most of them fairly weak and overlapping). At lower total pressure the spectral variations become amplified, since lines are broadened less (higher peaks and narrower widths). At the high temperatures common in combustion applications many more spectral lines appear, the so-called hot lines, generated by energy transition from molecules populating higher vibrational energy levels. For example, the HITEMP and CDSD databases [3, 4] contain about 36,000 lines for the spectral interval given in Fig. 1.2. The resulting absorption coefficient then resembles electronic noise [1]. This is clearly seen in Fig. 3.3, showing the pressure-based absorption coefficient of water vapor at 1000 K for a portion of its 6.3 μm band. Water vapor has millions of spectral lines covering most of the spectrum. HITEMP 2010 [3] contains the most important 111 million lines (other databases list up to 3 billion!), of which 9.5 million and 280,000, respectively, are in the spectral intervals shown in Fig. 3.3.

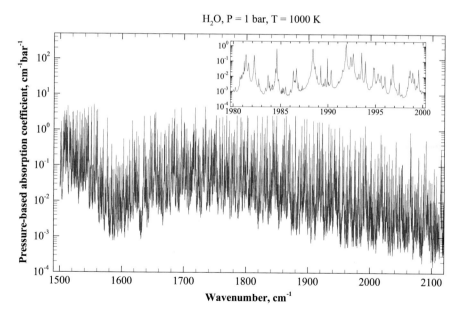

Fig. 3.3 Pressure-based spectral absorption coefficient for small amounts of H_2O in nitrogen; partial 6.3 μm band at $p = 1.0$ bar, $T = 1000$ K

Radiative Properties of Soot Soot particles are produced in fuel-rich parts of flames, as a result of incomplete combustion of hydrocarbon fuels (see discussion in Sect. 2.5.1). As shown by electron microscopy, soot particles are generally small and spherical, ranging in size between approximately 5 and 80 nm [5, 6]. While mostly spherical in shape, soot particles may also appear in agglomerated chunks and even as long agglomerated filaments. It has been determined experimentally in typical diffusion flames of hydrocarbon fuels that the volume fraction of soot, f_v, generally lies in the range between 10^{-5} and 10^{-8} [7–9].

Since soot particles are very small, they are generally at the same temperature as the flame and, therefore, strongly emit thermal radiation in a continuous spectrum over the infrared region. Experiments have shown that soot emission often is considerably stronger than the emission from the combustion gases. In order to predict the radiative properties of a soot cloud, it is necessary to determine the amount, shape, and distribution of soot particles, as well as their optical properties, which depend on chemical composition and particle porosity.

Early work on soot radiation properties concentrated on predicting the absorption coefficient κ_λ for a given flame as a function of wavelength λ. For all but the largest soot particles the size parameter $x = \pi d / \lambda$ (based on soot diameter d) is very small for all but the shortest wavelengths in the infrared, so one may expect that Rayleigh's theory for small particles will, at least approximately, hold. This condition would lead to negligible scattering and an absorption coefficient of

$$\kappa_\lambda = \beta_\lambda = -\Im\left\{\frac{m^2 - 1}{m^2 + 2}\right\}\frac{6\pi f_v}{\lambda} = \frac{36\pi nk}{(n^2 - k^2 + 2)^2 + 4n^2k^2}\frac{f_v}{\lambda}. \tag{3.21}$$

where β_λ is the extinction coefficient and $m = n - ik$ is the complex index of refraction. Experiments have confirmed that scattering may indeed be neglected [10]. It is customary to approximate the soot absorption coefficient by

$$\kappa_\lambda = \frac{C f_v}{\lambda^a}, \tag{3.22}$$

where C and a are empirical constants; values of the *dispersion exponent* a incorporate the spectral dependence of the complex index of refraction, ranging from 0.7 to as high as 2.2. However, the optical properties of soot material have also been measured directly by a number of experimenters. The most reliable ones today are perhaps those by Chang and Charalampopoulos [11] for propane soot, which have been corroborated by several other studies, and which have been curve-fit in polynomial form.

Agglomeration of soot into chunks or long chains renders the assumption of non-scattering soot questionable. The prediction of agglomeration requires complicated models, mostly due to Frenklach and coworkers [12–18]. The radiative properties of agglomerated soot have also been measured and modeled by a significant number of researchers, e.g., [19–23]. A brief review of such models and of how to estimate radiative properties of agglomerated soot has been given in the book by Modest [1].

3.3 Spectral Models

As indicated by the spectral variations of the absorption coefficient, such as that of CO_2 shown in Fig. 1.2, finding the radiative flux requires, in principle, the evaluation of the radiative transfer equation (RTE), Eq. (1.4), at an extremely large number of wavenumbers (perhaps one million), known as line-by-line calculations. Lack of computing resources and (until recently, before the advent of HITRAN and other databases) lack of accurate knowledge of absorption coefficient data have prompted the development of a number of "traditional" band models, in which averaged "line-of-sight" gas emissivities and transmissivities are estimated; as such they are limited to nonscattering media bounded by black walls. Modern band models (k-distributions) require high resolution databases; they reorder rather than average absorption coefficients, and can be applied to the RTE itself, i.e., they are valid for scattering media and/or reflecting walls.

3.3.1 The Optically Thin Approximation (OT)

In the *optically thin approximation* only the emission of radiation is considered, and it is assumed that all emission escapes from the flame without being absorbed, i.e., self-absorption is neglected. The mean free path a photon travels before absorption is equal to the inverse of the absorption coefficient, and the nondimensional product

$$\tau = \int_0^L \kappa \, ds \tag{3.23}$$

is known as the optical thickness of a medium, which may be interpreted as the ratio of distance over photon mean free path. Clearly, if $\tau \ll 1$ most photons will escape from the flame and self-absorption may be neglected. This simplifies the radiative source term given by Eq. (1.3) to

$$\dot{S}_{\text{rad}} = -\nabla \cdot \mathbf{q}_{\text{rad}} = -4\pi \int_0^\infty \kappa_\eta I_{b\eta} d\eta = -4\kappa_P \sigma T^4, \tag{3.24}$$

which is always a local heat loss (sink). The vast majority of combustion research papers to date that purported to consider "radiation" have used this gross simplification. The beauty of the OT is that it does not require the solution to any RTE, and it does predict the behavior of tiny laboratory flames fairly well. However, in larger flames for industrial applications the OT model overpredicts radiative heat loss by 50 % or more (and, thus, underpredicts flame temperatures by 100 °C or more).

3.3.2   Traditional Narrow Band Models

The formal solution to the spectral RTE for a nonscattering medium bounded by black walls along a straight line of sight is [1]

$$I_\eta(s) = I_{bw\eta} \exp\left(-\int_0^s \kappa_\eta ds'\right) + \int_0^s I_{b\eta}(s') \exp\left(-\int_{s'}^s \kappa_\eta ds''\right) \kappa_\eta(s')ds', \quad (3.25)$$

where I_η is the spectral intensity, $\eta = 1/\lambda$ is wavenumber (the spectral variable of choice for gas radiation), $I_{b\eta}$ and $I_{bw\eta}$ are the blackbody intensity of the medium and at the wall, respectively, and s is distance away from the wall. Assuming a homogeneous medium (constant temperature, pressure, and partial pressures of absorbing gases), i.e., $I_{b\eta}, \kappa_\eta = $ const, Eq. (3.25) may be written as

$$I_\eta(s) = I_{bw\eta} \tau_\eta(s) + I_{b\eta} \epsilon_\eta(s), \quad (3.26)$$

where

$$\tau_\eta(s) = 1 - \epsilon_\eta(s) = e^{-\kappa_\eta s} \quad (3.27)$$

are the transmissivity τ_η and emissivity ϵ_η, respectively, for a gas column of length s. Forming a narrow band average (with a $\Delta\eta = 4\text{--}25\,\text{cm}^{-1}$), and noting that the blackbody intensity essentially remains constant over such small spectral range, we obtain

$$\bar{I}_\eta = \frac{1}{\Delta\eta} \int_{\Delta\eta} I_\eta d\eta = \frac{1}{\Delta\eta} \int_{\Delta\eta} (I_{bw\eta}\tau_\eta + I_{b\eta}\epsilon_\eta)\, d\eta \simeq I_{bw\eta}\bar{\tau}_\eta + I_{b\eta}\bar{\epsilon}_\eta, \quad (3.28)$$

where

$$\bar{\tau}_\eta = \frac{1}{\Delta\eta} \int_{\Delta\eta} e^{-\kappa_\eta s} d\eta; \quad \bar{\epsilon}_\eta = \frac{1}{\Delta\eta} \int_{\Delta\eta} (1 - e^{-\kappa_\eta s})\, d\eta \quad (3.29)$$

are narrow band-averaged transmissivities and emissivities, respectively. We note from Figs. 1.2 and 3.3 that the absorption coefficient undergoes many oscillations across any narrow band, but that—if simple approximations for $\bar{\tau}_\eta$ and $\bar{\epsilon}_\eta$ can be found—the total intensity (or radiative flux) can be obtained in a (relatively) straightforward fashion from

$$I(s) = \int_0^\infty I_\eta(s) d\eta = \int_0^\infty \bar{I}_\eta(s) d\eta.$$

Several different narrow band models have been proposed, viz., the Elsasser model (assuming spectral lines to be of equal strength as well as equally spaced) and a number of statistical models (assuming different forms of randomness for line strength and spacing). While the Elsasser model is appropriate for diatomic molecules, it is recognized today that the Malkmus statistical model [1, 24] best represents multi-atomic combustion gases. In this model the placement of lines is random, while line strengths are picked from a probability distribution that accounts for the many weak lines that are always present. Using this model, the narrow band emissivity is evaluated from

$$\overline{\epsilon}_\eta(L) = 1 - \exp\left[-\frac{\beta}{2}\left(\sqrt{1 + \frac{4\tau}{\beta}} - 1\right)\right],\tag{3.30}$$

where L is the length of the gas column, β is a line overlap parameter, and τ (not to be confused with transmissivity τ_η) is the average optical depth, the latter two are defined by

$$\beta = \frac{\pi\gamma}{d}\;;\;\tau = \frac{S}{d}L,\tag{3.31}$$

where γ is the spectral lines' half-width at half-maximum (measured in cm^{-1}), d is the average line spacing (in cm^{-1}), and S is the average line strength (in cm^{-2}). It has been shown that for sufficiently small narrow bands ($\Delta\eta \leq 10\,cm^{-1}$) the Malkmus model can be optimized to predict transmissivities very accurately over wide ranges of parameters (mean error of better than 1 % for atmospheric or higher total pressures) [25]. Nonhomogeneous gas layers are somewhat problematical, but reasonable accuracy can be achieved with the so-called Curtis–Godson approximation [1] by defining path-averaged values for β and τ as

$$\tilde{\tau} = \int_0^L \frac{S}{d}\,dx\;,\;\tilde{\beta} = \frac{1}{\tilde{\tau}}\int_0^L \frac{S}{d}\beta\,dx.\tag{3.32}$$

Several databases are available for engineers wishing to use such narrow band models [26, 27]. The major limitation of traditional narrow band models is the fact that they can only predict line-of-sight transmissivities and emissivities, i.e., they cannot be incorporated into the RTE, precluding their use in scattering media and/or systems with reflecting surfaces.

3.3.3 Narrow Band k-Distributions

The realization that, under certain conditions, the spectrally oscillating absorption coefficient of a molecular gas can be reordered into a simpler, monotonically increasing function was first reported in the Western World some 30 years ago by atmospheric scientists, giving credit to earlier Russian work [28]. As for traditional

band models we will first look at the case of a homogeneous medium, i.e., a medium
with an absorption coefficient that, while varying across the spectrum, is spatially
constant. The RTE for such an emitting, absorbing, and scattering medium was
given in Eq. (1.4) as [1]

$$\frac{dI_\eta}{ds} = \kappa_\eta I_{b\eta} - (\kappa_\eta + \sigma_s)I_\eta + \frac{\sigma_s}{4\pi} \int_{4\pi} I_\eta(\hat{s}')\, \Phi(\hat{s}, \hat{s}')\, d\Omega'. \tag{3.33}$$

Let Eq. (3.33) be subject to the boundary condition at a wall

$$I_\eta = I_{w\eta} = \epsilon_w I_{bw\eta} + (1 - \epsilon_w)\frac{1}{\pi} \int_{\hat{n}\cdot\hat{s}<0} I_\eta\, |\hat{n}\cdot\hat{s}|\, d\Omega, \tag{3.34}$$

where $I_{w\eta}$ is the spectral intensity leaving the wall, ϵ_w is the wall's emittance, and
$\hat{n}$ is a unit surface normal pointing into the medium. In Eqs. (3.33) and (3.34)
we will assume that the spectral variation of scattering properties (σ_s, Φ) and surface
emittances (ϵ_w) is much more benign than that of the gas, i.e., we will assume
that these properties are constant (gray) across small parts of the spectrum (narrow
band). It is also true that across a narrow band (say, $\Delta\eta = 10\,\text{cm}^{-1}$) the Planck
functions $I_{b\eta}$ (medium) and $I_{bw\eta}$ (wall) remain essentially constant. Therefore, it
becomes clear that each location across the narrow band where the absorption
coefficient has one and the same value $\kappa_\eta = k$ will result in identical intensities I_η.
Recalculating the RTE each time would be rather wasteful; rather, the absorption
coefficient can be reordered into a monotonically increasing function, making
sure that a correct fraction of the narrow band contains an absorption coefficient
$k \leq \kappa_\eta \leq k + \delta k$, for all k, as indicated in Fig. 3.4a.

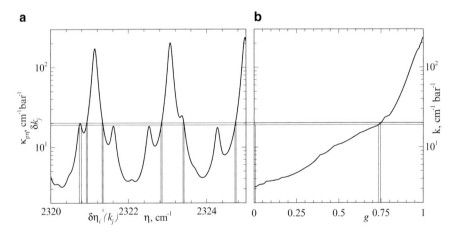

Fig. 3.4 Extraction of k-distributions from spectral absorption coefficient data (here for small
amounts of CO_2 in nitrogen, across a small part of its 4.3 μm band at $p = 1.0\,\text{bar}$, $T = 300\,\text{K}$);
(**a**) actual absorption coefficient and (**b**) reordered, equivalent k-distribution

Mathematically, this is achieved by multiplying Eqs. (3.33) and (3.34) with the Dirac delta function $\delta(k - \kappa_\eta)/\Delta\eta$, followed by integration of the narrow band. This leads to

$$\frac{dI_k}{ds} = kf(k)I_b - (k + \sigma_s)I_k + \frac{\sigma_s}{4\pi} \int_{4\pi} I_k(\hat{s}') \, \Phi(\hat{s}, \hat{s}') \, d\Omega' \tag{3.35}$$

with boundary condition

$$I_k = I_{wk} = \epsilon_w f(k)I_{bw} + (1 - \epsilon_w)\frac{1}{\pi} \int_{\hat{n}\cdot\hat{s}<0} I_k \, |\hat{n} \cdot \hat{s}| \, d\Omega, \tag{3.36}$$

where

$$I_k = \int_0^\infty I_\eta \, \delta(k - \kappa_\eta) \, d\eta \tag{3.37}$$

is the intensity I_η collected over all spectral locations where $\kappa_\eta = k$ (per dk), and

$$f(k) = \frac{1}{\Delta\eta} \int_{\Delta\eta} \delta(k - \kappa_\eta) \, d\eta \tag{3.38}$$

is known as the k-distribution, a probability density function for the absorption coefficient. While the narrow band-averaged intensity is readily obtained by integrating I_k over all values of k, the k-distributions tend to be quite poorly behaved, and the problem is further simplified by using the cumulative k-distribution, g,

$$g(k) = \int_0^k f(k) \, dk = \frac{1}{\Delta\eta} \int_{\Delta\eta} H(k - \kappa_\eta) \, d\eta, \tag{3.39}$$

where $H()$ denotes Heaviside's unit step function. Physically, $g(k)$ is the fraction of the narrow band over which $\kappa_\eta \leq k$. In practice, of course, the cumulative k-distribution is evaluated numerically, using k-bins of finite width δk, as shown in Fig. 3.4a for the j-th bin; A small step in g is then evaluated as

$$\delta g_j = f(k_j)\delta k_j = \frac{1}{\Delta\eta} \sum_i \left| \frac{\delta\eta}{\delta\kappa_\eta} \right|_i \left[H(k_j + \delta k_j - \kappa_\eta) - H(k_j - \kappa_\eta) \right]. \tag{3.40}$$

The redistributed absorption coefficient for the spectrum of Fig. 3.4a is shown in Fig. 3.4b. Note that both figures have identical maximum and minimum values for the absorption coefficient, and it is observed that absorption coefficients between 19 and 20 cm^{-1} (δk_j in Fig. 3.4a) occupy about 1.5 % of the narrow band spectrum (equal to the sum of $\delta\eta_j$ in Fig. 3.4a). Once the cumulative k-distribution has been found and inverted to yield $k(g)$, the "spectral" intensity I_g is found from the RTE as

$$\frac{dI_g}{ds} = k\left(I_{b\eta}(T) - I_g\right) - \sigma_s\left(I_g - \frac{1}{4\pi}\int_{4\pi} I_g(\hat{s}')\,\Phi(\hat{s}, \hat{s}')\,d\Omega'\right), \tag{3.41}$$

with the boundary conditions

$$I_g = I_{wg} = \epsilon_w I_{bw\eta} + (1 - \epsilon_w)\frac{1}{\pi}\int_{\hat{\mathbf{n}}\cdot\hat{s}<0} I_g\,|\hat{\mathbf{n}}\cdot\hat{s}|\,d\Omega, \tag{3.42}$$

where

$$I_g = I_k/f(k) = \frac{1}{\Delta\eta}\int_{\Delta\eta} I_\eta\,\delta\left(k - \kappa_\eta\right)d\eta \Big/ f(k), \tag{3.43}$$

and the narrow band-averaged intensity is evaluated from

$$\bar{I}_\eta = \frac{1}{\Delta\eta}\int_{\Delta\eta} I_\eta\,d\eta = \int_0^\infty I_k\,dk = \int_0^1 I_g\,dg. \tag{3.44}$$

As for the statistical models, application of the reordering concept to spatially nonhomogeneous absorption coefficients is somewhat problematical. It turns out the k-distribution approach is exact for a *correlated* absorption coefficient: at every wavenumber where $\kappa_\eta(\mathbf{r}_1)$ at one location has one and the same value, k, the absorption coefficient $\kappa_\eta(\mathbf{r}_2)$ at a different location always also has one unique value k^* (which may be a function of k but not η). If the ratio k^*/k is constant for all η across the narrow band (not a function of k) the absorption coefficient is *scaled*, i.e., spatial and spectral dependence are separable. Details on these restrictions are discussed in Modest [29]. As an illustration a simple (but severely nonhomogeneous) example is given in Fig. 3.5, showing transmissivities through a hot layer (50 cm width at 1000 K) adjacent to a cold slab (50 cm width at 300 K) of a 20 % H_2O–80 % N_2 mixture at a total pressure of 1 bar. Shown is the 6.3 μm vibration–rotation band of water vapor with a narrow band resolution of 25 cm^{-1}. Correspondence between exact LBL and k-distribution results is seen to be excellent (except for slight differences at a few wavenumbers) despite the severe nonisothermality. Recall that for isothermal media the k-distribution methods return exact transmissivities. An early very compact database of narrow band k-distributions for CO_2 and H_2O was collected by Soufiani and Taine [27]; a larger, high-accuracy database has been given by Wang and Modest [30] (and was recently extended and upgraded to HITEMP 2010 [31]).

3.3.4 Comparison of k-Distributions and Statistical Models

There are a number of important differences between the two types of band models, most favoring k-distributions:

Fig. 3.5 Narrow band
transmissivities for
two-temperature slab, as
calculated by the LBL,
scaled-*k*, and correlated-*k*
methods; 6.3 μm band of
H_2O with $p_{H_2O} = 0.2$ bar

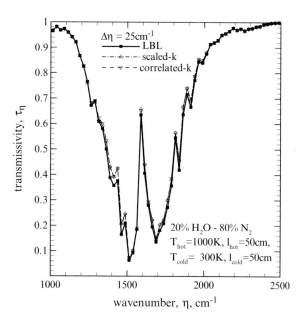

1. Statistical models are approximate for any gas while *k*-distributions are exact for homogeneous media.
2. Statistical models can only be used in nonscattering media bounded by black walls; *k*-distributions do not have either restriction.
3. For statistical models the allowable widths for a narrow band is determined by the gas spectrum's statistics ($\Delta\eta \lesssim 10\,\mathrm{cm}^{-1}$), while for *k*-distributions they are only limited by changes in the Planck function ($\Delta\eta \lesssim 100\,\mathrm{cm}^{-1}$).
4. However, if only a transmissivity is desired, the statistical models require just a single evaluation, while *k*-distributions require quadrature over the cumulative *k*-distributions *g*, perhaps ten RTE evaluations.
5. Both types of methods lose accuracy in nonhomogeneous media using different, reasonably successful approximate schemes.

3.4   Global Models

Global models attempt integration over the entire spectrum *before* solving the RTE, reducing RTE evaluations to a very small number. The first of these approaches was the weighted-sum-of-gray-gases (WSGG) model first presented by Hottel [32], designed for his then-popular zonal method. After Modest [33] showed it to be applicable to the general RTE, i.e., allowing its use with arbitrary RTE solution methods, the WSGG method became the most popular approach to deal with nongray combustion media. Further development by Denison and Webb [34–38] showed that the WSGG method, originally based on experimental total emissivity data, can also be used together with the new high-resolution databases, such as HITRAN and HITEMP; this variation is known as the spectral-line-based WSGG (SLW), or SLW method. A very similar approach, called the absorption distribution function method (ADF), was developed around the same time by Soufiani and coworkers [39–41]. Finally, Modest and coworkers [29, 42–45] have been able to extend the narrow band k-distribution concept to the entire spectrum, calling it the full-spectrum k-distribution method (FSK). It was found that the SLW/ADF approaches are low-level implementations of the FSK approach.

3.4.1 The Weighted-Sum-of-Gray-Gases Model

Revisiting Eq. (3.26), i.e., the symbolic solution to the RTE for a spatially constant absorption coefficient in a nonscattering medium, the equation can be integrated across the entire spectrum leading to

$$I(s) = \int_0^\infty I_\eta d\eta = I_{bw}(T_w)\left[1 - \epsilon(T_w, s)\right] + I_b(T)\epsilon(T, s), \qquad (3.45)$$

where

$$\epsilon(T, s) = \frac{1}{I_b(T)} \int_0^\infty (1 - e^{-\kappa_\eta s})\, I_{b\eta}(T)\, d\eta. \qquad (3.46)$$

Hottel assumed that this total emissivity of a gas column may be approximated by a weighted sum of gray gases, i.e.,

$$\epsilon(T, s) = \sum_{n=0}^N a_n(T)\left(1 - e^{-\kappa_n s}\right), \qquad (3.47)$$

where the gray-gas absorption coefficients κ_n are constants, while the weight factors a_n may be functions of source temperature; neither κ_n nor a_n are allowed to depend on path length s. Depending on the medium, the quality of the fit, and the accuracy desired, a N of 2 or 3 usually gives results of satisfactory accuracy [46]. Since, for an infinitely thick medium, the emissivity approaches unity, we find

$$\sum_{n=0}^N a_n(T) = 1. \qquad (3.48)$$

To accommodate pure molecular gases with their "spectral windows" (i.e., $\kappa_\eta \simeq 0$ between vibration–rotation bands) $\kappa_0 = 0$ by convention; the $n = 0$ term is simply dropped in the presence of absorbing particles. Sticking this into Eq. (3.45) leads to

$$I(s) = \sum_{n=0}^{N} I_n(s) = \sum_{n=0}^{N} \{[a_n I_b](T_w)e^{-\kappa_n s} + [a_n I_b](T)(1 - e^{-\kappa_n s})\}, \qquad (3.49)$$

i.e., each I_n is the solution to the RTE for a gray medium with absorption coefficient κ_n, but using a weighted Planck function $[a_n I_b]$. Early WSGG parameters for carbon dioxide–water vapor mixtures were based on now outdated experimental data [33, 47–49]. Recently, with the interest in oxy-fuel combustion and their higher content of radiating gases, a number of new correlations based on HITEMP [3] have been published [50–53]. All WSGG correlations are for fixed carbon dioxide–water vapor ratios, with the exception of Cassol et al. [53], which employs a mixing model similar to the more advanced SLW and FSK methods discussed in the next section.

Through the developments accompanying the SLW and FSK methods, it is known today that the WSGG may be applied also to scattering media and/or reflecting walls, i.e., Eqs. (3.33) and (3.34) may be used to find the I_n, simply by replacing κ_η and $I_{b\eta}(T)$ by κ_n and $[a_n I_b](T)$, respectively. Additionally, for any global model, it is assumed that scattering and surface properties (σ_s, Φ, ϵ_w) are gray. The greatest limitation of the WSGG method is its restriction to spatially constant absorption coefficients. No successful WSGG parameters for nonhomogeneous media appear to exist.

3.4.2 The Spectral-Line-Based WSGG Method

Hottel already made the physical argument that the weighted sum in Eq. (3.47) could also be interpreted as a nongray absorption coefficient field comprising N different values κ_k arbitrarily distributed across the spectrum, and occupying a fraction $a_k(T)$ of the Planck function weighted spectrum. Using this argument Denison and Webb [36, 54] collected *absorption line blackbody distribution functions* (ALBDF) for absorption cross-sections as

$$F_s(C_{abs}, T_b, T_g, p, x_s) = \frac{\pi}{\sigma T_b^4} \sum_i \int_{\Delta\eta \in (C_{abs,\eta}(T_g,p,x_s)<C_{abs})} I_{b\eta}(T_b)d\eta, \qquad (3.50)$$

where the sum is over all spectral regions $\Delta\eta$ for which $C_{abs,\eta}(T_g,p,x_s) < C_{abs}$. Here T_g is gas temperature, p is total pressure, x_s is mole fraction of species s, and the molar absorption cross-section is related to the absorption coefficient by

$$C_{abs,\eta} = \frac{R_u T_g}{p x_s}\kappa_\eta, \qquad (3.51)$$

with R_u being the universal gas constant. In a numerical calculation, the absorption cross-section domain is divided into discrete increments and a solution is found for each increment represented by a single value of the absorption cross-section or a single gray gas. For example, for a homogeneous gas with a single absorbing species, in the absence of scattering and within a black-walled enclosure one obtains

$$\frac{dI_i}{ds} = \tilde{k}_i(T)\left(\bar{a}_i(T)I_b(T) - I_i\right), \quad i = 1, \ldots, N,$$ (3.52)

for the N gray gases, subject to the boundary condition

$$I_{wi} = \bar{a}_i(T_w)I_b(T_w).$$ (3.53)

The weight function $\bar{a}_i$ in Eq. (3.52) is

$$\bar{a}_i(T) = F_s(C_{abs,i}, T, T, p, x_s) - F_s(C_{abs,i-1}, T, T, p, x_s),$$ (3.54)

while for the boundary condition $\bar{a}_i$ is evaluated as

$$\bar{a}_i(T_w) = F_s(C_{abs,i}, T_w, T, p, x_s) - F_s(C_{abs,i-1}, T_w, T, p, x_s).$$ (3.55)

Thus, $\bar{a}_i$ is the i-th finite range of the distribution function evaluated at the *local* Planck function temperature (T or T_w). The $\tilde{k}_i$ is an average value over the range $F_{i-1} < F_s \le F_i$. Denison and Webb [36, 54] suggest to evaluate $\tilde{k}_i$ from a square-root average, i.e.,

$$\tilde{k}_i(T) = \sqrt{k(T, F_{i-1})k(T, F_i)},$$ (3.56)

with k related to C_{abs} through Eq. (3.51). Inspection of Eq. (3.49) shows that the SLW scheme is simply the WSGG method, with absorption coefficients k_i and weights a_i evaluated from a line-by-line database. An example for a homogeneous CO_2–N_2 mixture is given in Fig. 3.6, with corresponding heat transfer results in Fig. 3.7. Results are given for SLW parameters calculated directly from the HITEMP 2010 database, and for parameters found from a simple correlation [55]; also included are LBL results, and values obtained with the full-spectrum k-distribution method of the following section. In this example $N = 5$ was chosen, which—as seen from Fig. 3.7—shows that the SLW (and, therefore, the WSGG method with properly chosen gray gases) results in respectable accuracy over a considerable range of optical thickness.

Denison and Webb also applied the SLW method to nonhomogeneous media, for which they deduced that the weight function $\bar{a}_i$ should be evaluated from

$$\bar{a}_i(T, \underline{\phi}_0) = F_s(C_{abs,i}, T, \underline{\phi}_0) - F_s(C_{abs,i-1}, T, \underline{\phi}_0),$$ (3.57)

where $\underline{\phi} = (T, p, x_s)$ is a vector defining the local gas state, and $\underline{\phi}_0 = (T_0, p_0, x_{s,0})$ is a reference state. Thus, $\bar{a}_i$ is the i-th finite range of the distribution function evaluated

Fig. 3.6 Planck function
weighted cumulative
k-distributions for 10 % CO_2
in nitrogen for gas and Planck
function temperatures of
1000 K, as evaluated from the
HITEMP database and the
correlation by Modest and
Mehta [55]; from [1]

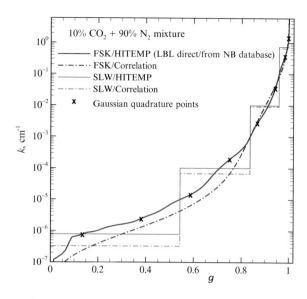

Fig. 3.7 Heat loss from an
isothermal slab of 10 % CO_2
in nitrogen at $T = 1000$ K, as
evaluated from the LBL,
FSK, and SLW models [1]

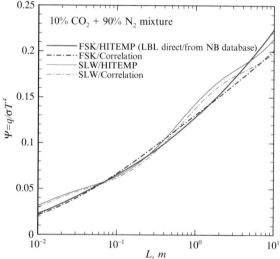

at the *local* Planck function temperature, while k is evaluated at the reference state.
The $\tilde{k}_i$ is calculated as

$$\tilde{k}_i(T_0, \underline{\phi}) = \sqrt{k_{i-1}(T_0, \underline{\phi})k_i(T_0, \underline{\phi})}, \qquad (3.58)$$

i.e., the value from the distribution function evaluated with the *local* absorption
coefficient and the Planck function evaluated at the *reference* temperature. These
rather complicated relationships for $\bar{a}$ and k_i were correctly deduced by Denison

and Webb [36], well before a solid theoretical foundation describing the interrela-
tionships between k-distributions was developed by Modest [29].

Rivière and coworkers employed the SLW method, but evaluated the gray-gas
parameters differently, calling it the ADF model [40]. The same group recognized
that for the case of strong temperature inhomogeneities one may separate a gas into
a number of "fictitious gases," grouping spectral lines according to the values of
their lower level energies, calling it the ADFFG (absorption distribution function
fictitious gases) method [41].

3.4.3 Full-Spectrum k-Distributions

The new mathematical description of narrow band k-distributions [29] as given by
Eqs. (3.35) and (3.36) paves the way for a reordering of the absorption coefficient
across the entire spectrum. Multiplying Eqs. (3.33) and (3.34) by the Dirac delta
function $\delta(k - \kappa_\eta)$, followed by integration over the entire spectrum, and assuming
scattering and surface properties (σ_s, Φ, ϵ_w) to be gray, leads to

$$\frac{dI_k}{ds} = kf(T,k)I_b - (k + \sigma_s)I_k + \frac{\sigma_s}{4\pi} \int_{4\pi} I_k(\hat{s}') \, \Phi(\hat{s}, \hat{s}') \, d\Omega' \tag{3.59}$$

with boundary condition

$$I_k = I_{wk} = \epsilon_w f(T_w, k)I_{bw} + (1 - \epsilon_w)\frac{1}{\pi} \int_{\hat{n}\cdot\hat{s}<0} I_k \, |\hat{n} \cdot \hat{s}| \, d\Omega, \tag{3.60}$$

where

$$I_k = \int_0^\infty I_\eta \, \delta(k - \kappa_\eta) \, d\eta \tag{3.61}$$

and

$$f(T,k) = \frac{1}{I_b} \int_0^\infty I_{b\eta}(T) \, \delta(k - \kappa_\eta) \, d\eta \tag{3.62}$$

is now a *Planck-function-weighted k-distribution*, which is a function of the gas
state at which the absorption coefficient is evaluated and of temperature T through
the Planck function. As for narrow band distributions, it is more desirable to cast the
RTE in terms of the cumulative k-distribution, now defined by

$$g(T,k) = \int_0^k f(T,k)dk = \frac{1}{I_b} \int_0^\infty I_{b\eta}(T)H(k - \kappa_\eta)d\eta. \tag{3.63}$$

In the same way as indicated in the discussion of the ALBDF in the SLW method, g is the fraction of the spectrum-integrated Planck function with an absorption coefficient $\kappa_\eta < k$. Therefore, $F(C_{abs})$ for SLW and $g(k)$ for FSK are mathematically identical (except for the use of C_{abs} instead of k in the SLW). Since the full-spectrum k-distribution is a function of temperature, one cannot simply divide Eqs. (3.59) and (3.60) by $f(T, k)$ as was done for narrow bands. Instead, one must define a reference temperature T_0, and the equations are divided by $f(T_0, k)$, leading to

$$\frac{dI_g}{ds} = k \left[a(T, T_0, g_0) I_b(T) - I_g \right] - \sigma_s \left(I_g - \frac{1}{4\pi} \int_{4\pi} I_g(\hat{s}') \, \Phi(\hat{s}, \hat{s}') \, d\Omega' \right), \quad (3.64)$$

subject to the boundary condition

$$I_g = I_{wg} = \epsilon_w a(T_w, T_0, g_0) I_{bw} + (1 - \epsilon_w) \frac{1}{\pi} \int_{\hat{n}\cdot\hat{s}<0} I_g \left| \hat{n} \cdot \hat{s} \right| d\Omega. \quad (3.65)$$

Here

$$I_g = I_k / f(T_0, k) = \int_0^\infty I_\eta \, \delta \left(k - \kappa_\eta \right) d\eta \Big/ f(T_0, k), \quad (3.66)$$

$$g_0(T_0, k) = \int_0^k f(T_0, k) \, dk, \quad (3.67)$$

$$a(T, T_0, g_0) = \frac{f(T, k)}{f(T_0, k)} = \frac{dg(T, k)}{dg_0(T_0, k)}, \quad (3.68)$$

and the total intensity is evaluated from

$$I = \int_0^\infty I_\eta \, d\eta = \int_0^\infty I_k \, dk = \int_0^1 I_g \, dg_0. \quad (3.69)$$

For homogeneous media the FSK method is exact, provided the integration in Eq. (3.69) is carried out accurately, e.g., using Gaussian quadrature with the quadrature points indicated in Fig. 3.6. Thus, the lines labeled LBL/FSK in Fig. 3.7 coincide. For nonhomogeneous media the equations remain valid provided the absorption coefficient is *correlated* or *scaled*. A reference state is chosen (based on total emission from the medium) where $\boldsymbol{\phi}_0 = (T_0, p_0, \mathbf{x}_0)$ is a vector containing all state variables influencing the absorption coefficient, such as temperature T, total pressure p, and species mole fractions $\mathbf{x}$. At this state the absorption coefficient is evaluated exactly and is used for the calculation of the reference k-distribution $f(T_0, k)$ and the stretching function $a(T, T_0, g_0)$. In the full-spectrum correlated-k (FSCK) method the absorption coefficient at other (nonreference) states is assumed to be correlated, and the k in Eq. (3.64) for a spatially invariable absorption coefficient is replaced by $k(T_0, \boldsymbol{\phi}, g_0)$, i.e., the k vs. g distribution found using the absorption coefficient evaluated at the local state ϕ, and the Planck function at the reference temperature T_0. In the full-spectrum scaled-k approach (FSSK) the absorption coefficient is assumed to be scaled, and k is replaced by $k(T_0, \phi_0, g_0)\, u(\phi, \phi_0)$, i.e., k vs. g evaluated at reference state and reference Planck function temperature, multiplied by a scaling function u. The latter approach tends to be somewhat more accurate, since the scaling function can be optimized; nonetheless it appears to be less popular since it is more difficult to apply. Details of both methods can be found in Modest [1, 29].

3.4.4   Full-Spectrum k-Distribution Assembly

Full-spectrum k-distributions (or, equivalently, ALBDFs) can provide answers essentially of LBL accuracy, but at a minuscule fraction of the computational cost (about 1:100,000). However, assembling these k-distributions as functions of Planck function temperature and state of the gas for every point in a three-dimensional enclosure is a tedious task at best. Thus, it is highly desirable to be able to retrieve them quickly from mathematical approximations or some kind of a database. Three different paths have been pursued:

Correlations   Using the hyperbolic tangent shape of a k-distribution as a base Denison and Webb and others generated mathematical correlations for atmospheric pressure for CO_2 and H_2O, requiring some 64 and 82 parameters, respectively, and employing now dated spectroscopic databases. More recent correlations have been given by Modest and Mehta [55], Modest and Singh [56], Liu et al. [57], and Pearson et al. [58], the last two being based on HITEMP 2010, and [58] also valid for elevated pressures. Since these correlations are for single species, approximate mixing schemes must be employed to generate k-distributions for gas (and soot) mixtures.

Assembly from a narrow band databases   Narrow band k-distributions do not depend on a Planck function temperature as a parameter and they may, of course, also be used for narrow band (as opposed to global) calculations. In addition, it was found that assembling mixture k-distributions is best carried out on a narrow band level [45]. An early compact (but now somewhat outdated) narrow band k-distribution database was given by Soufiani and Taine [27]; a larger, high-accuracy database has been given by Wang and Modest [30] (and was recently extended and upgraded to HITEMP 2010). The full-spectrum k-distribution is collected from narrow band data as

$$g(T, k) = \sum_{j \in \text{ all narrow bands}} \frac{I_{b\eta j}(T)}{I_b(T)} g_j(k), \qquad (3.70)$$

where $I_{b\eta j}$ is the Planck function $I_{b\eta}$ integrated across the j-th narrow band $\Delta \eta_j$.

Databases of full-spectrum k-distribution The facts that correlations have marginal accuracy, are fairly expensive to expensive to evaluate, and require expensive mixing (see below) have prompted the very recent generation of a comprehensive full-spectrum k-distribution for gas mixtures of H_2O, CO_2, and CO for a wide range of temperatures, mole fractions, and pressures [59]. To efficiently obtain accurate k-values [as well as the nongray stretching function a, Eq. (3.68)] for arbitrary thermodynamic states from tabulated values through a 6-D linear interpolation (3 mole fractions, temperature, pressure, and Planck function temperature) a very large database (5 GB in size) was generated. Using the concept of *dynamic loading* memory requirements and loading times are minimized for individual applications.

The correlations as well as the narrow band databases are for single species, and the assembly of k-distributions for a mixture requires a mixing model. The present full-spectrum k-distribution database also requires mixing if soot is to be considered. Several mixing models for k-distributions have been proposed, the two most popular (and accurate) ones being the *multiplicative scheme* and the *uncorrelated mixture scheme*.

Multiplicative Scheme In the multiplicative scheme of Solovjov and Webb [60] random line overlap is assumed, and the cumulative k-distribution of a mixture of M species is evaluated from

$$g_{mix}(T_P, T_g, p; k) = g_1(T_P, T_g, p; k) \times g_2(T_P, T_g, p; k) \times \cdots = \prod_{m=1}^{M} g_m(T_P, T_g, p; k). \tag{3.71}$$

Uncorrelated Mixture Scheme In the uncorrelated mixture scheme of Modest and Riazzi [45] it is assumed that lines from different species are uncorrelated, resulting in multiplicative transmissivities and a mixture cumulative k-distribution as

$$g_{mix}(k_{mix}) = \int_{g_1=0}^{1} \cdots \int_{g_M=0}^{1} H[k_{mix} - (k_1 + \cdots + k_M)]dg_M \ldots dg_1. \tag{3.72}$$

These mixing schemes can be applied at both the full spectrum and the narrow band level, but were found to be more accurate at the narrow band level. Soot is easily included at the narrow band level, since the soot absorption coefficient $\kappa_{soot,\eta}$ can be assumed constant across the j-th narrow band [but not across the entire spectrum, see Eq. (3.21)]. Then, for a soot-gas mixture

$$k_{mix}(g_j) = \kappa_{soot,j} + k_{gas}(g_j). \tag{3.73}$$

When mixing at the full-spectrum level soot must be taken as gray in order to use Eq. (3.73).

An example is given in the left frame of Fig. 3.8, showing the full-spectrum k-distributions for a mixture consisting of 12.5 % H_2O–15 % CO_2–72.5 % N_2 at $T = 1650$ K and $p = 1$ bar, i.e., values forcing interpolation for all correlations/databases

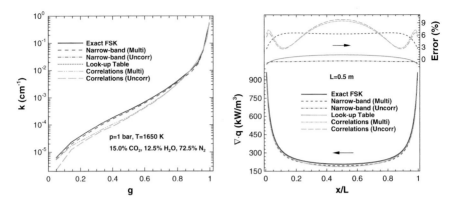

Fig. 3.8 k-distributions for a gas mixture (*left*) and corresponding divergence of radiative flux and errors compared to exact FSK for a homogeneous 1D slab (*right*) using different databases; 16-point quadrature

(except p, chosen as 1 bar because the correlations are limited to this single pressure). The right frame of Fig. 3.8 shows the radiative source, dq/dx, for such a mixture in a one-dimensional slab of $L = 50$ cm width, bounded by cold, black walls. The FSK method is exact for such a medium, the only errors stemming from (1) inaccuracy of the individual narrow band k-distributions, (2) errors due to interpolation, (3) inaccuracy of the mixing models, Eqs. (3.71) and (3.72), and (4) the neglected spectral variation of $I_{b\eta}$ across individual narrow bands. The accuracies of the look-up table and narrow band database with uncorrelated mixing are seen to be better than 1 %, i.e., comparable to the accuracy of the LBL calculations themselves. FSK calculations using the narrow band database with multiplicative mixing incurs up to 6 % error, while the correlations show a maximum error of almost 10 %.

Modest and Riazzi [45] also showed how a narrow band database can be employed to accurately and efficiently deal with the problem of nongray scattering and/or surface reflection by assembling part-spectrum k-distributions.

3.5 Radiative Transfer Equation Solution Methods

Several competing solution methods have been developed for the six-dimensional RTE, Eq. (1.4) (3 in space, 2 in direction, and a spectral variable), such as the zonal method, spherical harmonics method (SHM), discrete transfer method, discrete ordinates method (DOM), and the statistical Monte Carlo method (given in roughly historical sequence) [1]. Today the DOM and its modern cousin, the finite-volume method (DOM/FVM), are perhaps the most popular, and are implemented in some form in virtually all commercial CFD solvers. The SHM, a potentially more accurate cousin of the DOM, had not been seriously pursued by the research community until very recently, due to the daunting mathematics involved. Only its lowest order implementation, the P_1 method, is embedded into CFD solvers and enjoys enormous popularity, because it provides respectable accuracy at a very low computational cost (the solution to a single elliptic PDE). Photon Monte Carlo (PMC) are quickly becoming more and more popular, but are yet to be found in commercial and other open-source CFD codes.

The zonal method was developed in the 1950s by Hottel at MIT, as a semi-analytical tool to analyze a furnace broken up into a few "zones" and, before the development of other RTE methods, enjoyed great popularity. However, the method requires the inversion of a full matrix of exchange factors, making it computationally exceedingly expensive for modern grids (with millions of cells). This, and other shortcomings (inability to handle optically thick cells and anisotropic scattering) have essentially banished it from modern computations. Similarly, before the advent of the DOM the discrete transfer method enjoyed considerable popularity. The discrete transfer method is similar to the DOM inasmuch as discrete directions are chosen. It is also related to the Monte Carlo method (PMC), since rays of intensity are traced from surface to surface. Today it is recognized that this method essentially combines the disadvantages of, both, DOM and PMC, making it much slower than an equivalent DOM implementation [1]. Because of their established inferiority the zonal and discrete transfer methods will not be discussed, and we will rather concentrate on the more promising DOM/FVM, SHM, and PMC methods.

3.5.1 The Discrete Ordinates Method

The DOM (or S_N) method is a straightforward extension of the finite-difference method for spatial discretization, with additional discretization in the angular direction as well. In the DOM/FVM version, which is mostly employed today, both spatial and directional discretization are done in finite-volume fashion, transforming Eq. (1.4) for a given direction (or ordinate) $\hat{\mathbf{s}}_i$, and integrated over cell volume V to

$$\sum_k I_{ki}(\mathbf{s}_i \cdot \hat{\mathbf{n}}_k)A_k = \beta_p(S_{pi} - I_{pi})V\Omega_i, \tag{3.74a}$$

$$S_{pi} = (1 - \omega_p)I_{bp} + \frac{\omega_p}{4\pi}\sum_{j=1}^{n} I_{pj}\bar{\Phi}_{ij}, \tag{3.74b}$$

$$\bar{\Phi}_{ij} = \frac{1}{\Omega_i}\int_{\Omega_i}\int_{\Omega_j}\bar{\Phi}(\hat{\mathbf{s}}', \hat{\mathbf{s}})\,d\Omega'\,d\Omega, \tag{3.74c}$$

$$\mathbf{s}_i = \int_{\Omega_i}\hat{\mathbf{s}}\,d\Omega, \tag{3.74d}$$

where subscripts k and p imply evaluation at the center of the volume's faces A_k and element center P, respectively; subscript i denotes a value associated with solid angle Ω_i. The radiative source S_{pi}, combining emission and inscattering, has an analytically averaged phase function $\bar{\Phi}_{ij}$. Finally, the $\mathbf{s}_i$ is a vector (of varying length indicative of the size of Ω_i) pointing into an average direction within solid angle element Ω_i. The intensities at the face centers, I_{ki}, need to be related to those at volume centers, I_{pi}. There are many different ways to do this. However, the simple step scheme has generally been preferred, i.e., it is assumed that for intensities *leaving* control volume P (i.e., for $\mathbf{s}_i \cdot \hat{\mathbf{n}}_k > 0$) $I_{ki} = I_{pi}$. All *incoming* intensities ($\mathbf{s}_i \cdot \hat{\mathbf{n}}_k < 0$) are assigned the value of the element center from which they came. Substituting $I_{ki} = I_{pi}$ for $\mathbf{s}_i \cdot \hat{\mathbf{n}}_k > 0$ into Eq. (3.74) then leads to the final expression

$$I_{pi} = \frac{\beta_p S_{pi} V \Omega_i + \sum\limits_{k,in} I_{ki} |\mathbf{s}_i \cdot \hat{\mathbf{n}}_k| A_k}{\beta_p V \Omega_i + \sum\limits_{k,out} (\mathbf{s}_i \cdot \hat{\mathbf{n}}_k) A_k}, \tag{3.75}$$

where the "in" and "out" on the summation signs denote summation over volume faces with incoming ($\mathbf{s}_i \cdot \hat{\mathbf{n}}_k < 0$) or outgoing ($\mathbf{s}_i \cdot \hat{\mathbf{n}}_k > 0$) intensities, only.

Such formulation of first-order PDEs is very easily programmed, explaining its popularity. However, the method suffers from a few well-known shortcomings:

Ray effects   In problems with small, concentrated radiation sources, in geometries with large aspect ratios, or geometries with sharp corners (e.g., a wedge) and, most importantly, in grid systems with coarser directional than spatial resolution, the method produces erroneous results leading to so-called ray effects.

False scattering   False scattering is a consequence of spatial discretization errors, and is akin to "numerical diffusion" in CFD calculations. If a single, collimated beam is traced through an enclosure by the DOM, the beam will gradually widen as it moves farther away from its point of origin. This unphysical smearing of the radiative intensity, even in the absence of real scattering, is known as *false scattering* and can be reduced by using a finer mesh of control volumes (i.e., the opposite of what is required to alleviate ray effects).

Convergence   In the presence of scattering and/or wall reflections, the DOM equations are strongly coupled. In such a scenario, segregated solution of the directional equations (i.e., one direction at a time) results in poor convergence. The convergence of the DOM also deteriorates rapidly with increase in optical thickness of the medium. Numerically, this results in a poorly conditioned system of equations.

3.5.2 The Spherical Harmonics Method

The spherical harmonics P_N approximation is potentially more accurate than the DOM/FVM at comparable computational cost, but higher-order P_N are mathematically very complex and difficult to implement. The P_N method decouples spatial and directional dependencies by expanding the radiative intensity into a series of spherical harmonics. The lowest order of the P_N family, the P_1 approximation, has been extensively applied to radiative transfer problems. However, it loses accuracy when intensity is directionally very anisotropic [1], as is often the case in optically thin media. Applications of higher-order SHM methods were limited to one-dimensional cases for a long time, because of the cumbersome mathematics. Recently, Modest and Yang [61–63] have derived a general three-dimensional P_N formulation by eliminating odd spherical harmonics, consisting of $N(N+1)/2$ second-order elliptic PDEs and their Marshak boundary conditions for arbitrary geometries.

Expanding the radiative intensity into a sum of spherical harmonics leads to

$$I(\tau,\hat{s}) = \sum_{n=0}^{N} \sum_{m=-n}^{n} I_n^m(\tau)Y_n^m(\hat{s}), \tag{3.76}$$

where $\tau = \int \beta_r d\mathbf{r}$ is an optical coordinate, and β_r is the extinction coefficient. Y_n^m are the spherical harmonics,

$$Y_n^m = \begin{cases} \cos(m\psi)P_n^m(\cos\theta) & \text{for} \quad m \geq 0 \\ \sin(|m|\psi)P_n^m(\cos\theta) & \text{for} \quad m < 0 \end{cases} \tag{3.77}$$

where P_n^m are associated Legendre polynomials. The upper limit N in Eq. (3.76) is the order of the approximation. Substituting Eq. (3.76) into the RTE, and collecting spherical harmonics of individual order, leads to a system of $(N+1)^2$ simultaneous PDEs. After eliminating the I_n^m terms with odd n, a set of $N(N+1)/2$ elliptic PDEs [63] remains. The required boundary conditions are determined from general Marshak's conditions [64]. While these provide $N+1$ more equations than needed, Modest [63] recently showed that for the largest value of order N, only the even values of m should be employed for a consistent set of $N(N+1)/2$ boundary conditions.

The resulting lowest-order P_1-approximation for an isotropically scattering medium is easily stated as a single elliptic PDE in terms of the *incident radiation* $G = I_0^0$

$$\frac{1}{3\kappa}\nabla\cdot\left(\frac{1}{\beta}\nabla G\right) - G = -4\pi I_b, \tag{3.78}$$

with boundary condition

$$\mathbf{r} = \mathbf{r}_w: \qquad -\frac{2-\epsilon}{\epsilon}\frac{2}{3\beta}\hat{\mathbf{n}}\cdot\nabla G + G = 4\pi I_{bw}, \tag{3.79}$$

and

$$\mathbf{q} = -\frac{1}{3\beta}\nabla G. \tag{3.80}$$

The P_1-approximation is today implemented in virtually all commercial and open-source CFD solvers. Higher-order P_N-implementations exist today probably only in the authors' OpenFOAM [65] version. Similar to the DOM the SHM has a number of drawbacks:

P_1-limitations   While the simple P_1-approximation gives very respectable results for many situations at a very cheap computational price, it cannot provide accurate answers in fields with directionally strongly anisotropic intensities.

Mathematics/numerics   The mathematics for higher-order P_N are extremely involved, resulting in strongly coupled PDEs with cross-derivatives. Such equations are difficult to implement efficiently with standard tools available in CFD codes.

Directional limitations   In optically thin systems radiation intensities may exhibit abrupt directional variations, which are difficult to follow even by high-order spherical harmonic series (a similar situation as in the DOM's ray effects).

3.5.3   The Photon Monte Carlo Method

Many mathematical problems may also be solved by statistical methods, through sampling techniques, to any degree of accuracy. Such methods are often called Monte Carlo methods because of their use of random or chance events. There is no single scheme to which the name *Monte Carlo* applies. Rather, any method of solving a mathematical problem with an appropriate statistical sampling technique is commonly referred to as a Monte Carlo method; an example is the transported composition PDF solution scheme using stochastic particles, as described in Chap. 2.

Problems in thermal radiation are particularly well suited to solution by a Monte Carlo technique, since energy travels in discrete parcels (photons) over (usually) relatively long distances along a (usually) straight path before interaction with matter. Thus, solving a thermal radiation problem by Monte Carlo implies tracing the history of a statistically meaningful random sample of photons from their points of emission to their points of absorption. The advantage of the Monte Carlo method is that even the most complicated problem may be solved with relative ease. In particular, the method does not much care whether the medium is gray or strongly nongray, as is the case for combustion gases (for which up to 1 million spectral RTE solutions are needed if conventional RTE solvers are employed). The disadvantage of Monte Carlo methods is that, as statistical methods, they are subject to statistical error (very similar to the unavoidable error associated with experimental measurements). This can sometimes be problematic if the remainder of the model is deterministic (such as RANS solutions of the flow field), but is certainly not an issue when combined with other stochastic methods, such as the transported composition PDF. In order to distinguish the methods from another we will here use the name PMC for the radiation model.

In order to follow the history of radiative energy bundles in a statistically meaningful way, the locations, directions and wavenumbers of emission, reflective behavior of surfaces, etc., must be chosen according to probability distributions. Emission points of rays are determined randomly according to the emissive energy distribution in the domain. In hot zones of combustion gases the concentrations of carbon dioxide and water vapor tend to be high and strong emission is observed. Thus, hot zones must be represented by more rays than cold zones with weak emission, and the ray number density reflects the emissive energy distribution, which is achieved by constructing random number relations for emission locations [1]. By inverting such random number relations, the emission location can be determined as

$$x = x(R_x), \quad y = y(R_y, x), \quad z = z(R_z, x, y), \tag{3.81}$$

where R_x, R_y, and R_z are independent random numbers. Since emission is isotropic, similar, very simple relations are found for statistical emission directions [1].

The determination of a statistically meaningful wavenumber of emission for a nongray combustion gas is a little more involved, and will serve as an example here. The probability of the number of photons emitted in a differential wavenumber interval $d\eta$ is proportional to the Planck function weighted by the spectral absorption coefficient, i.e.,

$$\text{Probability}\{\eta \text{ in } d\eta\} \propto \kappa_\eta I_{b\eta}\, d\eta. \tag{3.82}$$

Therefore, to simulate this emission process of photons statistically for a given gas species i, the random number relation for the emission wavenumber is derived as

$$R_\eta = \frac{\int_0^\eta \kappa_{\eta,i} I_{b\eta}\, d\eta}{\int_0^\infty \kappa_{\eta,i} I_{b\eta}\, d\eta} = \frac{\int_0^\eta \kappa_{p\eta,i} I_{b\eta}\, d\eta}{\int_0^\infty \kappa_{p\eta,i} I_{b\eta}\, d\eta} = \frac{\pi}{\kappa_{p,i}\sigma T^4} \int_0^\eta \kappa_{p\eta,i} I_{b\eta}\, d\eta, \tag{3.83}$$

where R_η is a random number uniformly distributed in $[0, 1)$, $\kappa_{p\eta,i} = \kappa_{\eta,i}/p_i$ is the pressure-based spectral absorption coefficient and p_i is the partial pressure of species i; $\kappa_{p,i}$ is the pressure-based Planck-mean absorption coefficient. In practice one deals with gas mixtures. Considering that the absorption coefficient is additive,

$$\kappa_\eta = \sum_i \kappa_{\eta,i} = \sum_i \kappa_{p\eta,i} p_i \quad \text{and} \quad \kappa_{p\eta} = \kappa_\eta/p = \sum_i x_i \kappa_{p\eta,i}, \tag{3.84}$$

where $x_i = p_i/p$ is the mole fraction of species i and p is the total pressure of the mixture, one can obtain the random number relation for the gas mixture from Eq. (3.83) as

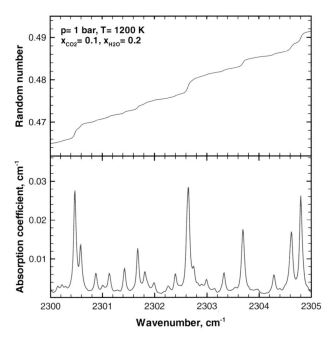

Fig. 3.9 Random number and absorption coefficient distributions in a small spectral interval (absorption coefficient data obtained from Wang and Modest [66])

$$R_\eta = \frac{\pi}{\kappa_p \sigma T^4} \int_0^\eta \kappa_{p\eta} I_{b\eta}\, d\eta = \frac{\pi}{\sigma T^4 \sum_i x_i \kappa_{p,i}} \sum_i x_i \int_0^\eta \kappa_{p\eta,i} I_{b\eta}\, d\eta. \qquad (3.85)$$

Equation (3.85) establishes a direct relation between the mixture random number R_η and wavenumber of emission η. However, this relationship is very difficult to invert, but also impossible to database because the infinite number of gas mixture possibilities. Figure 3.9 shows the random number and corresponding absorption coefficient distributions of a gas mixture in a small spectral interval. Although the random number is a monotonically increasing function, it has strongly varying gradients even in such a small interval. A small error in random number may result in a significant deviation in absorption coefficient. Therefore, common root-finding techniques relying on smooth gradients, such as the Newton Raphson method, cannot be used here to invert random numbers. A number of investigations have been dedicated to make this determination of emission wavenumbers accurate and efficient [67–69].

Photon bundles emitted from a given location, into a given direction, and at a given wavenumber are then traced through the computational domain, where they may be absorbed or scattered by other molecules during transmission and may be reflected or absorbed by bounding walls. The bundles carry a fixed amount of energy and the optical thickness they travel through is determined randomly as

$$\tau_{abs} = \ln(1/R_\tau), \quad \tau_{abs} = \int_0^{s^*} \kappa \, ds, \tag{3.86}$$

where τ_{abs} is the predetermined optical thickness for a specific ray, R_τ is the random number uniformly distributed in $[0,1)$, s is the position variable along the path, and s^* is the location at which the ray is absorbed. Starting from the emission location, the optical thickness passed through by a ray is accumulated during tracing until its corresponding τ_o value is reached, then all the energy carried by the ray is absorbed by the subvolume at the absorbing location. This method is called the *standard* or *ballistic* Monte Carlo method. If the medium is optically thin, most of the rays may exit open boundaries without contributing to the statistics in the medium. Similarly, if the medium is optically very thick, rays travel only very short distances without contributing to statistics of elements far away. In both cases the standard method becomes very inefficient. An alternative, the *energy partitioning* method, was proposed to alleviate such problem [70, 71], in which the ray energy is absorbed gradually and partitioned to all the subvolumes it traverses until depletion or exiting.

PMC for Stochastic Media Turbulent combustion models use stochastic probability density function (PDF) models to resolve the nonlinear turbulence–chemistry interaction (TCI) term, in which the fluid is represented by a large number of notional point masses (see discussion in Chap. 2). To simulate the radiative transfer process by ray tracing in a discrete particle field, the interaction between infinitesimal point masses and infinitesimally thin photon rays needs to be modeled. This results in different emission and absorption algorithms (from and interacting with individual point particles rather than throughout the volume). This can be done by assigning effective volumes to the point masses, by assigning an influence volume to the ray's trajectory, or a combination of both. Several particle models and ray models were developed by Wang and Modest [72, 73], as well as photon emission and absorption algorithms based on these models. In the cone–PPM scheme the particles are treated as point masses and the rays as narrow cones. The only geometric information known about the particles is their position vector $\mathbf{r}_i$. However, particles do have a nominal volume V_i, which may be calculated from their thermophysical properties. The absorption coefficient in a discrete particle field is represented by a set of Dirac delta functions,

$$\kappa = \sum_i \kappa_i V_i \delta(\mathbf{r} - \mathbf{r}_i). \tag{3.87}$$

If the ray is modeled as a cone, it can interact with point particles by "capturing" them. The optical distance such a cone of rays traverses can then be calculated as [72]

$$\tau = \sum_{i \in I} \frac{\kappa_i V_i}{\pi R_{c,i}^2} = \sum_{i \in I} \frac{\kappa_{\rho,i} m_i}{\pi R_{c,i}^2}, \tag{3.88}$$

where I denotes all the particles enclosed by the cone and $R_{c,i}$ is the radius of the cone at the location of the ith particle.

3.6 Turbulence–Radiation Interactions

We noted in Eq. (2.1) how the radiative source as defined by Eq. (1.3) enters the overall energy equation, in the same way mass sources enter into the species equations. In the RANS context, if a composition PDF is employed, all composition variables are determined from the PDF, and the radiative source term appears on the right-hand side of Eq. (2.24), as well as its Lagrangian equivalent Eq. (2.28), i.e., it cannot be expressed in terms of local composition variables and must be modeled. S_{rad} comprises two parts: absorption of incoming radiation and emission from the volume itself,

$$S_{\text{rad}} = S_{\text{abs}} - S_{\text{emi}}, \tag{3.89}$$

where

$$S_{\text{emi}} = 4\pi \int_0^\infty \kappa_\eta I_{b\eta} d\eta = 4\pi \kappa_P(T, p_\alpha) I_b(T) \tag{3.90}$$

depends entirely on local scalars (temperature, partial pressures of radiating species) and does not need to be modeled. On the other hand,

$$S_{\text{abs}} = \int_0^\infty \kappa_\eta \int_{4\pi} I_\eta d\Omega d\eta = \int_0^\infty \kappa_\eta G_\eta d\eta \tag{3.91}$$

incorporates the radiative intensity, which is a nonlocal quantity and, through the RTE, depends on composition variables everywhere within the medium (as well as their local fluctuations). Its mean, therefore, must be modeled if a conventional RTE approach is used.

Taking the Reynolds average of Eq. (3.89) brings the turbulence–radiation interactions (TRI) into evidence:

Emission TRI : $\langle S_{\text{emi}} \rangle = 4\pi \langle \kappa_P I_b \rangle = 4\pi \langle \kappa_P \rangle \langle I_b \rangle + \langle \kappa_P' I_b' \rangle,$ (3.92a)

Absorption TRI : $\langle S_{\text{abs}} \rangle = \int_0^\infty \langle \kappa_\eta G_\eta \rangle d\eta = \int_0^\infty \left[\langle \kappa_\eta \rangle \langle G_\eta \rangle + \langle \kappa_\eta' G_\eta' \rangle \right] d\eta.$
(3.92b)

These TRI terms contain several nonlinearities, which are sometimes useful to examine separately, such as the

Absorption coefficient self-correlation:

$$\mathscr{R}_\kappa = \kappa_P(\langle T \rangle, \langle p_\alpha \rangle) / \langle \kappa_P(T, p_\alpha) \rangle, \tag{3.93a}$$

Temperature self-correlation:

$$\mathscr{R}_{I_b} = I_b(\langle T \rangle) / \langle I_b(T) \rangle = \langle T \rangle^4 / \langle T^4 \rangle, \tag{3.93b}$$

Absorption coefficient–Planck function correlation:

$$\mathscr{R}_{\kappa I_b} = \langle \kappa_P \rangle \langle I_b \rangle) / \langle \kappa_P I_b \rangle, \tag{3.93c}$$

Emission correlation:

$$\mathscr{R}_{\text{emiss}} = \kappa_P(\langle T \rangle, \langle p_\alpha \rangle) I_b(\langle T \rangle) / \langle \kappa_P I_b \rangle = \mathscr{R}_\kappa \mathscr{R}_{I_b} \mathscr{R}_{\kappa I_b}. \tag{3.93d}$$

Since radiative emission only depends on local properties, this allows an independent analysis of the character of emission TRI without considering turbulent reactive flow details. Coelho [74] carried out such an analysis for turbulent diffusion flames of methane. The instantaneous thermochemical state of the reactive mixture is described by a flamelet model using the GRI 3.0 mechanism [75] and determined from the numerical solution of the flamelet equations. These equations yield the temperature and the chemical composition of the reactive mixture as a function of mixture fraction and scalar dissipation rate, in the absence of turbulent fluctuations. The Flame Master code [76] was used to solve the flamelet equations for scalar dissipation rates, χ, ranging from near equilibrium to near flame quenching conditions. The pdf shape of mixture fraction was assumed to be a beta function. Calculations were performed for diffusion flames of methane burning in air at atmospheric pressure, for a scalar dissipation rate $\chi = 1\,\text{s}^{-1}$ and a radiant fraction of $\chi_R = 10\,\%$, given that turbulent scalar fluctuations effects on radiative emission are essentially independent of the scalar dissipation rate and radiant fraction. Figure 3.10 shows the values of $\mathscr{R}_\kappa$, $\mathscr{R}_{I_b}$, $\mathscr{R}_{\kappa I_b}$, and $\mathscr{R}_{\text{emiss}}$ as a function of mixture fraction for different levels of turbulent fluctuations. The dashed vertical line indicates stoichiometry conditions ($\bar{z} = 0.055$), with lean conditions (as prevalent over most regions of a diffusion flame) to the left, and fuel-rich conditions to the right. It is observed that over lean regions of the flame turbulence significantly reduces the mean absorption coefficient ($\mathscr{R}_\kappa > 1$) while strongly increasing the mean Planck function ($\mathscr{R}_{I_b} < 1$). On the other hand, the interaction between absorption coefficient and Planck function (a measure of the nongrayness of the medium) can be both positively and negatively correlated. Total emission from the flame, Fig. 3.10d, is always increased by turbulence ($\mathscr{R}_{\text{emiss}} < 1$), as consistently observed by all experimental and theoretical investigations.

Absorption TRI is much harder to model, since it depends on composition variables and their fluctuations everywhere within the medium. However, Kabashnikov and Myasnikova [77] as well as Song and Viskanta [78] recognized that the fluctuations of local intensity are governed by points far away and are only weakly correlated with the fluctuations of the local absorption coefficient, if the mean free path for radiation is much larger than the turbulence length scale. Thus, within the framework of the *optically thin eddy approximation* or *optically thin fluctuation assumption* (OTFA), we have

$$\langle \kappa_\eta G_\eta \rangle \simeq \langle \kappa_\eta \rangle \langle G_\eta \rangle \quad \text{if} \quad \kappa_\eta l_t \ll 1, \tag{3.94}$$

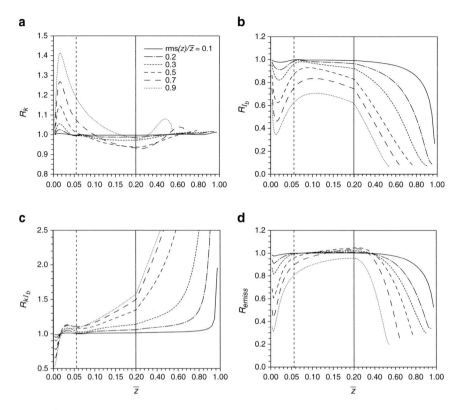

Fig. 3.10 Influence of turbulent scalar fluctuations on (**a**) the mean absorption coefficient, (**b**) Planck function, (**c**) absorption coefficient–correlation, and (**d**) radiative emission, as a function of mean and rms of mixture fraction [74]

where l_t is an appropriate turbulence length scale. The *absorption coefficient self-correlation* $\langle \kappa_\eta \rangle$ is still a nonlinear average but depends only on local quantities and does not need modeling. The validity of OTFA depends on eddy size distribution and the absorption coefficient of the mixture. Hartick et al. [79] have shown that in combustion gases OTFA may not be valid over very small parts of the spectrum (e.g., parts of the strong 4.3 μm CO_2 band), but that their effect on overall results is negligible. Similarly, Coelho [80] showed OTFA to be valid for the modeling of Sandia Flame D, and Wang et al. [81] demonstrated that this is also true for larger nonsooting flames. On the other hand, in strongly sooting flames the absorption coefficient may be very large over substantial parts of the spectrum and OTFA may be violated. Invoking the OTFA, i.e., neglecting absorption TRI, the mean radiation source may be evaluated as

$$\langle S_{\mathrm{rad}} \rangle \simeq \int_0^\infty \left(\langle \kappa_\eta \rangle \langle G_\eta \rangle - 4\pi \langle \kappa_\eta I_{b,\eta} \rangle \right) d\eta. \tag{3.95}$$

If the FSK method of Sect. 3.4.3 together with Gaussian quadrature for spectral integration is employed Eq. (3.95) reduces to

$$\langle S_{\text{rad}} \rangle \simeq \sum_{j=1}^{M} w_j \langle k_j \rangle \langle G_j \rangle - 4\pi \sum_{j=1}^{M} w_j \langle k_j a_j I_b \rangle, \tag{3.96}$$

where M is the total number of quadrature points and the w_j are the quadrature weights. The PDF (2.24), and its Lagrangian counterpart Eq. (2.28), may then be restated as

$$\frac{\partial \rho f_\phi}{\partial t} + \frac{\partial \rho \tilde{u}_i f_\phi}{\partial x_i} + \frac{\partial \rho S_\alpha f_\phi}{\partial \psi_\alpha} + \delta_{\alpha(h)} 4\pi \sum_{j=1}^{M} w_j \frac{\partial (k_j a_j I_b) f_\phi}{\partial \psi_\alpha}$$

$$= -\frac{\partial}{\partial x_i}[\langle u_i'' | \boldsymbol{\psi} \rangle \rho f_\phi] + \frac{\partial}{\partial \psi_\alpha}[\langle \frac{\partial J_i^\alpha}{\partial x_i} | \boldsymbol{\psi} \rangle f_\phi]$$

$$+ \delta_{\alpha(h)} \sum_{j=1}^{M} w_j \frac{\partial (k_j \langle G_j \rangle) f_\phi}{\partial \psi_\alpha} \tag{3.97}$$

$$dx_i^* = \tilde{u}_i^* dt + \left(\langle \rho \rangle^{-1} \frac{\partial \Gamma_{T\phi}}{\partial x_i} \right)^* dt + \left(2 \langle \rho \rangle^{-1} \Gamma_{T\phi} \right)^{*1/2} dW_i,$$

$$d\phi_\alpha^* = S_\alpha(\boldsymbol{\phi}^*) dt - \frac{1}{2} C_\phi (\phi_\alpha^* - \tilde{\phi}_\alpha) \omega dt$$

$$+ \delta_{\alpha(h)} \sum_{j=1}^{M} w_j \left[(k_j \langle G_j \rangle) - 4\pi (k_j a_j I_b) \right] \rho^{-1}(\boldsymbol{\psi}) dt. \tag{3.98}$$

If absorption TRI is to be accounted for, such as in strongly sooting flames, S_{abs} must be modeled to greater accuracy, or must be calculated directly. The first attempt to quantify absorption TRI was made by Tessé and coworkers [82], who investigated a small sooting (luminous) ethylene flame, using detailed chemistry and a sophisticated soot model [83], together with a Lagrangian solver to obtain the composition PDF. They then constructed many homogeneous turbulence structures from this PDF and determined the thermal radiation with a PMC scheme together with the narrow band k-distribution model of Soufiani and Taine [27]. They found emission to increase by 30 %, and also found absorption TRI to be appreciable (5 % of total emission) for this luminous flame, indicating eddies of appreciable optical thickness. Other approximate (and probably unsatisfactory) schemes were offered by Mehta and Modest [84] (applying the radiative diffusion approximation to optically thick eddies) and Coelho [85] (using a two-dimensional clipped Gaussian joint probability density function of mixture fraction and radiation intensity). The first ones to assess absorption TRI from basic principles (i.e., without the assumptions for turbulence structures made by Tessé) were Wang and coworkers [81], who also used a transported composition PDF to determine composition variables and their turbulence moments, but together with Wang's [67, 73] LBL-accurate PMC scheme for stochastic particles. This radiation solver was specifically developed to determine a PDF for photons, providing full compatibility with the stochastic turbulence model. With their model Wang et al. [81] provided proof that absorption TRI is negligible for Sandia Flame D and, indeed, also for large nonluminous flames. The method was further employed to investigate the influence of TRI in sooting flames by Mehta et al. [86–88], who modeled six sooting flames [89–91] using Wang et al.'s schemes together with a sophisticated soot model [92]. In contrast to Tessé's observations, absorption TRI was found to be negligible for all six laboratory-scale flames, despite the soot. Only when scaling up the sootiest flame [90] by a factor of 32 did absorption TRI become appreciable (6 % of total emission).

In the LES framework, TRI are brought into evidence by spatially filtering Eq. (3.89) [using Eqs. (2.18) and (2.19)] to yield:
Emission TRI:

$$\langle S_{\text{emi}} \rangle_\Delta = 4\pi \langle \kappa_P I_b \rangle_\Delta \tag{3.99a}$$

$$= 4\pi \left[\langle \langle \kappa_P \rangle_\Delta \langle I_b \rangle_\Delta \rangle_\Delta + \langle \langle \kappa_P \rangle_\Delta I_b' \rangle_\Delta + \langle \kappa_P' \langle I_b \rangle_\Delta \rangle_\Delta + \langle \kappa_P' I_b' \rangle_\Delta \right],$$

Absorption TRI:

$$\langle S_{\text{abs}} \rangle_\Delta = \int_0^\infty \langle \kappa_\eta G_\eta \rangle_\Delta d\eta \tag{3.99b}$$

$$= \int_0^\infty \left[\langle \langle \kappa_\eta \rangle_\Delta \langle G_\eta \rangle_\Delta \rangle_\Delta + \langle \langle \kappa_\eta \rangle_\Delta G_\eta' \rangle_\Delta + \langle \kappa_\eta' \langle G_\eta \rangle_\Delta \rangle_\Delta + \langle \kappa_\eta' G_\eta' \rangle_\Delta \right] d\eta.$$

As was the case for TCI (Sect. 2.4), there are additional TRI terms in LES compared to RANS. The four terms on the right-hand side of Eq. (3.99a) correspond, respectively, to the correlation between resolved-scale fluctuations in κ_P and resolved-scale fluctuations in I_b, the correlation between resolved-scale fluctuations in κ_P and subfilter-scale fluctuations in I_b, the correlation between subfilter-scale fluctuations in κ_P and resolved-scale fluctuations in I_b, and the correlation between subfilter-scale fluctuations in κ_P and subfilter-scale fluctuations in I_b. Only the first term is captured explicitly in LES; the other three contributions must be modeled, and the relative magnitude of each term will vary with the filter size Δ. For example, as Δ decreases, a larger fraction of the fluctuations will be captured explicitly and the contributions of the subfilter-scale fluctuations will decrease. In the case of a LES/PDF method where κ_P and I_b are known functions of the composition variables [e.g., Eq. (2.30) in the case of a particle-based implementation], all contributions to emission TRI are captured and no further modeling is needed.

A similar interpretation applies to the four terms on the right-hand side of Eq. (3.99b), with κ_η replacing κ_P and G_η replacing I_b. As discussed earlier in the RANS context, absorption TRI are more difficult because of the nonlocal nature of G_η. The OTFA for correlations involving subfilter-scale fluctuations is expected to be an even better approximation in LES compared to RANS, and the approximation should improve with decreasing filter size Δ (decreasing local optical thickness based on the filter scale). Also, absorption TRI can be closed using the particle-based LES/PDF framework that was described earlier for RANS/PDF, with the appropriate changes in physical interpretation and scale specification as discussed in Sects. 2.3 and 2.4.

Further discussion of the individual TRI terms in LES can be found in [93–96]. It is of interest to understand the relative importance of the various emission and absorption TRI terms, and how those vary with flame conditions and with the LES filter scale Δ, so that appropriate approximations and models can be developed. That can be done using DNS data (where all fluctuations are fully resolved) or results from LES/PDF modeling studies (where all TRI contributions are included—albeit in the context of a model) that are analyzed to isolate and quantify the relative contributions of resolved-scale fluctuations and subfilter-scale fluctuations. Both DNS and LES/PDF studies are discussed in the examples that are included in subsequent chapters.

References

1. M.F. Modest, *Radiative Heat Transfer*, 3rd edn. (Academic, New York, 2013)
2. L.S. Rothman, I.E. Gordon, A. Barbe, D.C. Benner, P.F. Bernath, M. Birk, V. Boudon, L.R. Brown, A. Campargue, J.-P. Champion, K. Chance, L.H. Coudert, V. Dana, V.M. Devi, S. Fally, J.-M. Flaud, R.R. Gamache, A. Goldman, D. Jacquemart, I. Kleiner, N. Lacome, W.J. Lafferty, J.-Y. Mandin, S.T. Massie, S.N. Mikhailenko, C.E. Miller, N. Moazzen-Ahmadi, O.V. Naumenko, A.V. Nikitin, J. Orphal, V.I. Perevalov, A. Perrin, A. Predoi-Cross, C.P. Rinsland, M. Rotger, M. Simeckova, M.A.H. Smith, K. Sung, S.A. Tashkun, J. Tennyson, R.A. Toth, A.C. Vandaele, J.V. Auwera, The HITRAN 2008 molecular spectroscopic database. J. Quant. Spectrosc. Radiat. Transf. **110**, 533–572 (2009)
3. L.S. Rothman, I.E. Gordon, R.J. Barber, H. Dothe, R.R. Gamache, A. Goldman, V.I. Perevalov, S.A. Tashkun, J. Tennyson, HITEMP, the high-temperature molecular spectroscopic database. J. Quant. Spectrosc. Radiat. Transfer **111**(15), 2139–2150 (2010)
4. S.A. Tashkun, V.I. Perevalov, CDSD-4000: high-resolution, high-temperature carbon dioxide spectroscopic databank, J. Quant. Spectrosc. Radiat. Transf. **112**(9), 1403–1410 (2011). Available from ftp://ftp.iao.ru/pub/CDSD4000
5. J.M. Singer, J. Grumer, Carbon formation in very rich hydrocarbon–air flames—I: studies of chemical content, temperature, ionization and particulate matter, in *Seventh Symposium (International) on Combustion* (The Combustion Institute, Pittsburg, PA, 1959), pp. 559–572
6. B.L. Wersborg, J.B. Howard, G.C. Williams, Physical mechanisms in carbon formation in flames, in *Fourteenth Symposium (International) on Combustion* (The Combustion Institute, Pittsburg, PA, 1972), pp. 929–940
7. A.F. Sarofim, H.C. Hottel, Radiative transfer in combustion chambers: influence of alternative fuels, in *Proceedings of the Sixth International Heat Transfer Conference*, vol. 6 (Hemisphere, Washington, DC, 1978), pp. 199–217
8. M. Kunugi, H. Jinno, Determination of size and concentration of soot particles in diffusion flames by a light-scattering technique, in *Eleventh Symposium (International) on Combustion* (The Combustion Institute, Pittsburg, PA, 1966), pp. 257–266
9. T. Sato, T. Kunitomo, S. Yoshi, T. Hashimoto, On the monochromatic distribution of the radiation from the luminous flame. Bull. JSME **12**, 1135–1143 (1969)

10. A. Becker, Über die Strahlung und Temperatur der Hefnerlampe. Ann. Phys. **333**(5), 1017–1031 (1909)
11. H. Chang, T.T. Charalampopoulos, Determination of the wavelength dependence of refractive indices of flame soot, Proc. R. Soc. (Lond.) A **430**(1880), 577–591 (1990)
12. M. Frenklach, H. Wang, M. J. Rabinowitz, Optimization and analysis of large chemical kinetic mechanisms using the solution mapping method—combustion of methane. Progr. Energy Combust. Sci. **18**, 47–73 (1992)
13. M. Frenklach, H. Wang, Detailed mechanism and modeling of soot particle formation, in *Soot Formation in Combustion* (Springer, New York, 1994), pp. 162–192
14. M. Frenklach, On surface growth mechanism of soot particles, in *Twenty-Sixth Symposium (International) on Combustion* (The Combustion Institute, Pittsburg, PA, 1996), pp. 2285–2293
15. H. Wang, M. Frenklach, A detailed kinetic modeling study of aromatics formation in laminar premixed acetylene and ethylene flames. Combust. Flame **110**, 173–221 (1997)
16. M. Frenklach, S.J. Harris, Aerosol dynamics modeling using the method of moments. J. Colloid Interface Sci. **118**, 252–261 (1987)
17. M. Frenklach, S.J. Harris, Aerosol dynamics using the method of moments, J. Colloid Interface Sci. **130**, 252–261 (1987)
18. M. Frenklach, Soot Modeling Home Page, http://www.me.berkeley.edu/soot
19. R.A. Dobbins, C.M. Megaridis, Morphology of flame-generated soot as determined by thermophoretic sampling. Langmuir **3**, 254–259 (1987)
20. M.F. Iskander, H.Y. Chen, J.E. Penner, Optical scattering and absorption by branched-chains of aerosols. Appl. Opt. **28**, 3083–3091 (1989)
21. Ü.Ö. Köylü, G.M. Faeth, Radiative properties of flame-generated soot. J. Heat Transf. **115**(2), 409–417 (1993)
22. S. Manickavasagam, M.P. Mengüç, Scattering matrix elements of fractal-like soot agglomerates. J. Appl. Phys. **36**(6), 1337–1351 (1997)
23. T.L. Farias, M.G. Carvalho, Ü.Ö. Köylü, G.M. Faeth, Computational evaluation of approximate Rayleigh–Debye–Gans fractal-aggregate theory for the absorption and scattering properties of soot. J. Heat Transf. **117**(1), 152–159 (1995)
24. W. Malkmus, Random Lorentz band model with exponential-tailed S^{-1} line-intensity distribution function, J. Opt. Soc. Am. **57**(3), 323–329 (1967)
25. A.A. Lacis, V. Oinas, A description of the correlated-k distribution method for modeling nongray-gaseous absorption, thermal emission, and multiple scattering in vertically inhomogeneous atmospheres. J. Geophys. Res. **96**(D5), 9027–9063 (1991)
26. W.L. Grosshandler, RADCAL: a narrow-band model for radiation calculations in a combustion environment. Technical Report NIST Technical Note 1402, National Institute of Standards and Technology (1993)
27. A. Soufiani, J. Taine, High temperature gas radiative property parameters of statistical narrow-band model for H_2O, CO_2 and CO, and correlated-k model for H_2O and CO_2. Int. J. Heat Mass Transf. **40**(4), 987–991 (1997)
28. A. Arking, K. Grossman, The influence of line shape and band structure on temperatures in planetary atmospheres. J. Atmos. Sci. **29**, 937–949 (1972)
29. M.F. Modest, Narrow-band and full-spectrum k-distributions for radiative heat transfer—correlated-k vs. scaling approximation. J. Quant. Spectrosc. Radiat. Transf. **76**(1), 69–83 (2003)
30. A. Wang, M.F. Modest, High-accuracy, compact database of narrow-band k-distributions for water vapor and carbon dioxide, in *Proceedings of the ICHMT 4th International Symposium on Radiative Transfer*, Istanbul, Turkey, 2004, ed. by M.P. Mengüç, N. Selçuk
31. J. Cai, M.F. Modest, Improved full-spectrum k-distribution implementation for inhomogeneous media using a narrow-band database. J. Quant. Spectrosc. Radiat. Transf. **141**, 65–72 (2013)
32. H.C. Hottel, A.F. Sarofim, *Radiative Transfer* (McGraw-Hill, New York, 1967)
33. M.F. Modest, The weighted-sum-of-gray-gases model for arbitrary solution methods in radiative transfer. J. Heat Transf. **113**(3), 650–656 (1991)

34. M.K. Denison, B.W. Webb, A spectral line based weighted-sum-of-gray-gases model for arbitrary RTE solvers. J. Heat Transf. **115**, 1004–1012 (1993)

35. M.K. Denison, B.W. Webb, k-Distributions and weighted-sum-of-gray gases: a hybrid model, in *Tenth International Heat Transfer Conference* (Taylor & Francis, London, 1994), pp. 19–24.

36. M.K. Denison, B.W. Webb, The spectral-line-based weighted-sum-of-gray-gases model in nonisothermal nonhomogeneous media. J. Heat Transf. **117**, 359–365 (1995)

37. M.K. Denison, B.W. Webb, Development and application of an absorption line blackbody distribution function for CO_2. Int. J. Heat Mass Transf. **38**, 1813–1821 (1995)

38. M.K. Denison, B.W. Webb, The spectral-line weighted-sum-of-gray-gases model for H_2O/CO_2 mixtures. J. Heat Transf. **117**, 788–792 (1995)

39. P. Rivière, A. Soufiani, M.-Y. Perrin, H. Riad, A. Gleizes, Air mixture radiative property modelling in the temperature range 10000–40000 K. J. Quant. Spectrosc. Radiat. Transf. **56**, 29–45 (1996)

40. L. Pierrot, A. Soufiani, J. Taine, Accuracy of narrow-band and global models for radiative transfer in H_2O, CO_2, and $H_2O–CO_2$ mixtures at high temperature. J. Quant. Spectrosc. Radiat. Transf. **62**, 523–548 (1999)

41. L. Pierrot, P. Rivière, A. Soufiani, J. Taine, A fictitious-gas-based absorption distribution function global model for radiative transfer in hot gases. J. Quant. Spectrosc. Radiat. Transf. **62**, 609–624 (1999)

42. M.F. Modest, H. Zhang, The full-spectrum correlated-k distribution and its relationship to the weighted-sum-of-gray-gases method, in *Proceedings of the IMECE 2000*, vol. HTD-366-1 (American Society of Mechanical Engineers, Orlando, FL, 2000), pp. 75–84

43. H. Zhang, M.F. Modest, A multi-scale full-spectrum correlated-k distribution for radiative heat transfer in inhomogeneous gas mixtures. J. Quant. Spectrosc. Radiat. Transf. **73**(2–5), 349–360 (2002)

44. H. Zhang, M.F. Modest. Scalable multi-group full-spectrum correlated-k distributions for radiative heat transfer. J. Heat Transfer **125**(3), 454–461 (2003)

45. M.F. Modest, R.J. Riazzi, Assembly of full-spectrum k-distributions from a narrow-band database; effects of mixing gases, gases and nongray absorbing particles, and mixtures with nongray scatterers in nongray enclosures. J. Quant. Spectrosc. Radiat. Transf. **90**(2), 169–189 (2005)

46. H.C. Hottel, A.F. Sarofim, *Radiative Transfer* (McGraw-Hill, New York, 1967)

47. T.F. Smith, Z.F. Shen, J.N. Friedman, Evaluation of coefficients for the weighted sum of gray gases model. J. Heat Transf. **104**, 602–608 (1982)

48. I.H. Farag, T.A. Allam, Gray-gas approximation of carbon dioxide standard emissivity. J. Heat Transf. **103**, 403–405 (1981)

49. J.S. Truelove, The zone method for radiative heat transfer calculations in cylindrical geometries. HTFS Design Report DR33 (Part I: AERE-R8167) (Atomic Energy Authority, Harwell, 1975)

50. T. Kangwanpongpan, F.H.R. França, R.C. da Silva, P.S. Schneider, H.J. Krautz, New correlations for the weighted-sum-of-gray-gases model in oxy-fuel conditions based on HITEMP 2010 database. Int. J. Heat Mass Transf. **55**, 7419–7433 (2012)

51. L.J. Dorigon, G. Duciak, R. Brittes, F. Cassol, M. Galarça, F.H.R. França, WSGG correlations based on HITEMP2010 for computation of thermal radiation in non-isothermal, non-homogeneous H_2O/CO_2 mixtures. Int. J. Heat Mass Transf. **64**, 863–873 (2013)

52. M.H. Bordbar, G. Wecel, T. Hyppänen, A line by line based weighted sum of gray gases model for inhomogeneous $CO_2–H_2O$ mixture in oxy-fired combustion. Combust. Flame **161**, 2435–2445 (2014)

53. F. Cassol, R. Brittes, F.H.R. França, O.A. Ezekoye, Application of the weighted-sum-of-gray-gases model for media composed of arbitrary concentrations of H_2O, CO_2 and soot. Int. J. Heat Mass Transf. **79**, 796–806 (2014)

54. M.K. Denison, B.W. Webb, An absorption-line blackbody distribution function for efficient calculation of total gas radiative transfer. J. Quant. Spectrosc. Radiat. Transf. **50**, 499–510 (1993)

55. M.F. Modest, R.S. Mehta, Full spectrum k-distribution correlations for CO_2 from the CDSD-1000 spectroscopic databank. Int. J. Heat Mass Transf. **47**, 2487–2491 (2004)
56. M.F. Modest, V. Singh, Engineering correlations for full spectrum k-distribution of H_2O from the HITEMP spectroscopic databank. J. Quant. Spectrosc. Radiat. Transf. **93**, 263–271 (2005)
57. F. Liu, H. Chu, H. Zhou, G.J. Smallwood. Evaluation of the absorption line blackbody distribution function of CO_2 and H_2O using the proper orthogonal decomposition and hyperbolic correlations. J. Quant. Spectrosc. Radiat. Transf. **128**, 27–33 (2013)
58. J.T. Pearson, B.W. Webb, V.P. Solovjov, J. Ma, Efficient representation of the absorption line blackbody distribution function for H_2O, CO_2, and CO at variable temperature, mole fraction, and total pressure. J. Quant. Spectrosc. Radiat. Transf. **138**, 82–96 (2014)
59. C. Wang, W. Ge, M.F. Modest, J. Cai, A full-spectrum k-distribution look-up table for radiative transfer in nonhomogeneous gaseous media, in *CHT-15: Advances in Computational Heat Transfer* (Begell House, Redding, CT, 2015)
60. V.P. Solovjov, B.W. Webb, SLW modeling of radiative transfer in multicomponent gas mixtures. J. Quant. Spectrosc. Radiat. Transf. **65**, 655–672 (2000)
61. M.F. Modest, J. Yang, Elliptic PDE formulation and boundary conditions of the spherical harmonics method of arbitrary order for general three-dimensional geometries. J. Quant. Spectrosc. Radiat. Transf. **109**, 1641–1666 (2008)
62. J. Yang, M.F. Modest, High-order *P-N* approximation for radiative transfer in arbitrary geometries. J. Quant. Spectrosc. Radiat. Transf. **104**(2), 217–227 (2007)
63. M.F. Modest, Further developments of the elliptic P_N-approximation formulation and its Marshak boundary conditions. Numer. Heat Transfer B **62**(2–3), 181–202 (2012)
64. R.E. Marshak, Note on the spherical harmonics method as applied to the Milne problem for a sphere. Phys. Rev. **71**, 443–446 (1947)
65. H. Jasak, A. Jemcov, Z. Tukovic, OpenFOAM: A C++ library for complex physics simulations, in *International Workshop on Coupled Methods in Numerical Dynamics* (Inter-University Centre, Dubrovnik, 2007), pp. 1–20
66. A. Wang, M.F. Modest, Importance of combined Lorentz–Doppler broadening in high-temperature radiative heat transfer applications. J. Heat Transf. **126**(5), 858–861 (2004)
67. A. Wang, M.F. Modest, Spectral Monte Carlo models for nongray radiation analyses in inhomogeneous participating media. Int. J. Heat Mass Transf. **50**, 3877–3889 (2007)
68. A.M. Feldick, M.F. Modest, An improved wavelength selection scheme for Monte Carlo solvers applied to hypersonic plasmas. J. Quant. Spectrosc. Radiat. Transf. **112**, 1394–1401 (2011)
69. T. Ren, M.F. Modest. Hybrid wavenumber selection scheme for line-by-line photon Monte Carlo simulations in high-temperature gases. J. Heat Transfer **135**(8), 084501 (2013)
70. M.F. Modest, S.C. Poon, Determination of three-dimensional radiative exchange factors for the space shuttle by Monte Carlo. ASME paper no. 77-HT-49 (1977)
71. M.F. Modest, Determination of radiative exchange factors for three dimensional geometries with nonideal surface properties. Numer. Heat Transf. **1**, 403–416 (1978)
72. A. Wang, M.F. Modest, Photon Monte Carlo simulation for radiative transfer in gaseous media represented by discrete particle fields. J. Heat Transf. **128**, 1041–1049 (2006)
73. A. Wang, M.F. Modest, An adaptive emission model for Monte Carlo ray-tracing in participating media represented by statistical particle fields. J. Quant. Spectrosc. Radiat. Transf. **104**(2), 288–296 (2007)
74. P.J. Coelho, A theoretical analysis of the influence of turbulence on radiative emission in turbulent diffusion flames of methane. Combust. Flame **160**, 610–617 (2013)
75. G.P. Smith, D.M. Golden, M. Frenklach, N.W. Moriarty, B. Eiteneer, M. Goldenberg, C.T. Bowman, R. Hanson, S. Song, W.C. Gardiner, V. Lissianski, Z. Qin, GRI-Mech 3.0 (1999). Available at http://www.me.berkeley.edu/gri_mech
76. H. Pitsch, FlameMaster v3.3.10, A C++ Computer Program for 0D Combustion and 1D Laminar Flame Calculations (2015). Available at http://www.itv.rwth-aachen.de/index.php?id=128

77. V.P. Kabashinikov, G. I. Myasnikova, Thermal radiation in turbulent flows—temperature and concentration fluctuations. Heat Transf.-Sov. Res. **17**(6), 116–125 (1985)
78. T.-H. Song, R. Viskanta, Interaction of radiation with turbulence: application to a combustion system. J. Thermophys. Heat Transf. **1**(1), 56–62 (1987)
79. J.W. Hartick, M. Tacke, G. Fruchtel, E.P. Hassel, J. Janicka, Interaction of turbulence and radiation in confined diffusion flames, in *Twenty-Sixth Symposium (International) on Combustion* (The Combustion Institute, Pittsburg, PA, 1996), pp. 75–82
80. P.J. Coelho, Detailed numerical simulation of radiative transfer in a nonluminous turbulent jet diffusion flame. Combust. Flame **136**, 481–492 (2004)
81. A. Wang, M.F. Modest, D.C. Haworth, L. Wang, Monte Carlo simulation of radiative heat transfer and turbulence interactions in methane/air jet flames. J. Quant. Spectrosc. Radiat. Transf. **109**(2), 269–279 (2008)
82. L. Tessé, F. Dupoirieux, J. Taine, Monte Carlo modeling of radiative transfer in a turbulent sooty flame. Int. J. Heat Mass Transf. **47**, 555–572 (2004)
83. B. Zamuner, F. Dupoirieux, Numerical simulation of soot formation in a turbulent flame with a Monte-Carlo PDF approach and detailed chemistry. Combust. Sci. Technol. **158**, 407–438 (2000)
84. R.S. Mehta, M.F. Modest, Modeling absorption TRI in optically thick eddies, in *Proceedings of Eurotherm Seminar 78*, April (Elsevier, Poitiers, 2006)
85. P.J. Coelho, Assessment of a presumed joint pdf for the simulation of turbulence–radiation interaction in turbulent reactive flows. Appl. Therm. Eng. **49**, 22–30 (2012)
86. R.S. Mehta, A. Wang, M.F. Modest, D.C. Haworth, Modeling of a turbulent ethylene/air flame using hybrid finite volume/Monte Carlo methods. Comput. Therm. Sci. **1**, 37–53 (2009)
87. R.S. Mehta, D.C. Haworth, M.F. Modest, Composition PDF/photon Monte Carlo modeling of moderately sooting turbulent jet flames. Combust. Flame **157**, 982–994 (2010)
88. R.S. Mehta, M.F. Modest, D.C. Haworth, Radiation characteristics and turbulence–radiation interactions in sooting turbulent jet flames. Combust. Theor. Model. **14**(1), 105–124 (2010)
89. A. Coppalle, D. Joyeux, Temperature and soot volume fraction in turbulent diffusion flames: measurements of mean and fluctuating values. Combust. Flame **96**, 275–285 (1994)
90. J.H. Kent, D. Honnery, Modeling sooting turbulent jet flames using an extended flamelet technique. Combust. Sci. Technol. **54**, 383–397 (1987)
91. N.E. Endrud. Soot, radiation and pollutant emissions in oxygen-enhanced turbulent jet flames. Master's thesis, The Pennsylvania State University, University Park, PA, 2000
92. R.S. Mehta, D.C. Haworth, M.F. Modest, An assessment of gas-phase thermochemistry and soot models for laminar atmospheric-pressure ethylene–air flames. Proc. Combust. Inst. **32**, 1327–1334 (2009)
93. M. Roger, C.B.D. Silva, P.J. Coelho, Analysis of the turbulence–radiation interactions for large eddy simulations of turbulent flows. Int. J. Heat Mass Transf. **52**, 2243–2254 (2009)
94. C.B. da Silva, I. Malico, P.J. Coelho, Radiation statistics in homogeneous isotropic turbulence. New J. Phys. **11**, 093001-1–34 (2009)
95. M. Roger, P.J. Coelho, C.B. da Silva, The influence of the non-resolved scales of thermal radiation in large eddy simulation of turbulent flows: a fundamental study. Int. J. Heat Mass Transf. **53**, 2897–2907 (2010)
96. M. Roger, P.J. Coelho, C.B. da Silva, Relevance of the subgrid-scales for large eddy simulations of turbulence-radiation interactions in a turbulent plane jet. J. Quant. Spectrosc. Radiat. Transf. **112**, 1250–1256 (2011)

Chapter 4
Radiation Effects in Laminar Flames

Even in the absence of turbulence, radiative heat transfer has important influences on the global and local behavior of flames. There is a large and rapidly growing body of literature containing analytic, experimental and simulation studies aimed at isolating and quantifying the influences of radiation on (for example) laminar flame speeds and pollutant emissions in laminar flames. Radiation plays an especially prominent role in "threshold" phenomena, including flammability, extinction, and stability limits for laminar flames. These effects can be amplified in turbulent flames. In this chapter, several examples of important radiation effects in laminar and transitional flames are discussed, before returning to the fully turbulent flames that are the primary focus of this monograph. Spatially one-dimensional systems are considered first, followed by two-dimensional systems.

4.1 One-Dimensional Laminar Flames

Premixed flames, where the reactants (fuel and oxidizer) are mixed at the molecular level prior to initiation of flame propagation, are considered first. This is followed by a discussion of radiation effects in nonpremixed flames, or diffusion flames, where fuel and oxidizer are segregated initially and combustion occurs as the reactants mix at the molecular level.

4.1.1 1D Laminar Premixed Flames

The speed at which a laminar flame propagates through a homogeneous premixed reactant mixture (the *laminar flame speed* or *laminar burning velocity*) is an

© The Author(s) 2016
M.F. Modest, D.C. Haworth, *Radiative Heat Transfer in Turbulent Combustion Systems*, SpringerBriefs in Applied Sciences and Technology,
DOI 10.1007/978-3-319-27291-7_4

important fundamental property of a reactive mixture. It is also of interest to quantify the thermochemical conditions under which steady laminar flame propagation can be sustained (the *flammability limits*) for a premixed reactant mixture. The influences of radiative heat transfer had been largely neglected in early simplified analyses of premixed laminar flame propagation and in the interpretation of experimental results, but radiation effects have been considered in a number of more recent analytic, numerical, and experimental studies.

Flammability limits and radiation-induced extinction of stretched premixed laminar flames, including nonunity Lewis number effects (different molecular transport rates for species and for energy), were analyzed for a counterflow premixed twin-flame configuration using large-activation-energy asymptotics in [1]. There explicit expressions for the stretch and radiation extinction limits and flammability limits were derived, and those were found to be in good qualitative agreement with results from experiments and numerical simulations.

In a numerical study of one-dimensional (planar) premixed laminar flames [2], it was found that reabsorption of emitted radiation led to substantially higher burning velocities and wider extinction limits than in calculations using an optically thin radiation model, particularly when CO_2 was present in the unburned gas. It was further concluded that fundamental, configuration-independent flammability limits could exist due to radiative heat losses, and that the limits were strongly dependent on the emission–absorption spectra of the reactant and product gases and their temperature dependencies. An example is shown in Fig. 4.1. There the results with consideration of reabsorption lie between the adiabatic (no radiation) and optically thin results for methane–air flames, whereas when CO_2 is present in the unburned gas, the peak temperatures and laminar burning velocities with consideration of reabsorption are higher than those for the adiabatic case. For a similar configuration, a theoretical/numerical analysis of flame propagation through a particle-gas mixture representing an idealized dust cloud [3] showed that the mixture temperature

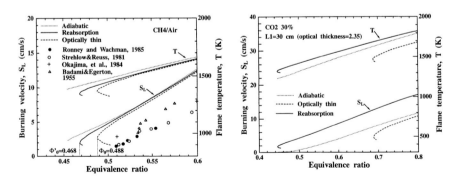

Fig. 4.1 Computed laminar burning velocities and peak temperatures as functions of equivalence ratio neglecting radiation ("Adiabatic"), using an optically thin radiation model ("Optically thin"), and using a radiation model that includes reabsorption ("Reabsorption") [2]. *Left*: lean methane–air mixtures. *Right*: lean methane–air mixture with 30 % CO_2

through the preheat zone increased due to thermal radiation, suggesting that radiation could play a significant role in two-phase flame propagation.

Steady planar one-dimensional laminar premixed flames are difficult to realize experimentally, and transient spherical laminar flames (a spherical flame propagating through a homogeneous reactant mixture from the ignition point) have been used extensively to deduce laminar flame speeds and flammability/extinction limits. Motivated by discrepancies between measured and computed laminar flame speeds in very lean mixtures, a computational study of propagating spherical flames in constant-volume vessels was performed using different radiation models in [4]. It was found that neglect of radiation and compression effects could result in significant under-prediction of the laminar flame speeds that are extracted from propagating spherical flame studies. Uncertainties in flame speed measurements in this configuration due to radiation were explored analytically and numerically in [5]. There an analytical estimate of the effects of radiation heat loss was derived and validated against detailed numerical simulations. The results showed that flames with low adiabatic flame speeds (close to the extinction limits) could be strongly inhibited by radiative heat losses, especially at early times (Fig. 4.2). Uncertainties associated with the extraction of laminar flame speeds from experimental measurements in spherically expanding flames and in counterflow premixed flames were assessed using detailed one-dimensional simulations in [6], focusing on effects of molecular transport and radiation. Again, it was found that radiation losses affected the extracted laminar flame speed mainly for slower flames. They also reported that the uncertainty due to radiation could be eliminated by directly measuring the velocity in the unburned gas immediately in front of the flame using (for example) high-speed particle-image velocimetry. An empirical correlation quantifying the uncertainty in laminar flame speed associated with radiation in spherically expanding premixed flames was reported in [7]. The correlation was shown to work for different fuels at ambient and elevated temperatures and pressures, and a method

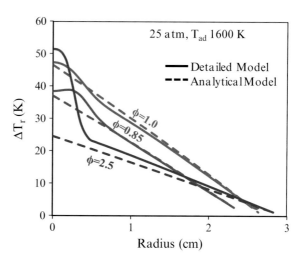

Fig. 4.2 Detailed and analytical model results for burned-gas temperature differences with versus without radiation for 25-atm flames with He–Ar dilution at three equivalence ratios [5]

to obtain a "radiation-corrected flame speed" was presented. Extensions to consider radiation effects for two-phase flows (dust clouds) in the spherical premixed flame configuration have been reported in numerical studies [8] and in analytic studies [9].

Radiation effects also have been explored in other one-dimensional premixed laminar flame configurations. This includes an analysis of premixed tubular flames and premixed planar counterflow flames [10], where the effects of flame curvature, radiation, and stretch on flame extinction were investigated using the large-activation-energy asymptotic method with a nonlinear radiation model. A general expression for flame speed, flame temperature, and extinction limits was obtained, and was used to study the radiation and flame curvature coupling for different Lewis numbers. The results showed that the coupling between radiation and flame curvature leads to multiple flame bifurcations and extinction limits. Both the stretch-induced and the radiation-induced extinction limits were found to be affected by flame curvature. A one-dimensional numerical model was used to generate a stabilization diagram for methane/air premixed flames in a porous-medium foam with a uniform ambient temperature in [11], including radiation effects. And the impact of radiative heat transfer on the behavior of flat-flame burners (a planar premixed flame stabilized downstream of, or within, a porous plug) was assessed using a simplified one-dimensional model in [12]. There the porous plug was modeled as a thermally conducting, optically thick medium, allowing for both conductive and radiative heat transfer. The results showed that radiative heat losses played a critical role in two different combustion modes. In "surface" combustion, the flame stabilized at a stand-off distance from the porous plug that was determined by heat transfer between the gas phase and the porous solid; there radiation served as a mode of heat dissipation to prevent flash-back, where the flame moves upstream into the porous matrix. In "submerged" combustion, the flame stabilized within the porous plug itself. There radiation prevented flame stabilization close to the inlet and exit faces, and enabled a "slow" solution branch that did not exist without consideration of radiative losses.

4.1.2   1D Laminar Diffusion Flames

In laminar nonpremixed (diffusion) flames, extinction limits are of primary importance. Hydrodynamic-stretch-induced extinction (where a flame is extinguished by high local velocity gradients) has been of primary interest in the combustion literature, but radiation-induced extinction (where radiative cooling results in extinction) can also occur. An early analysis showing that radiative extinction could occur in laminar diffusion flames at low stretch rates was reported in [13]. Large-activation-energy asymptotics were used to investigate oscillatory instabilities near the radiation-induced extinction limit at large and small Damköhler numbers in [14].

Counterflow diffusion flames, where a fuel jet impinges against an oxidizer jet resulting in a quasi-one-dimensional system along the stagnation line, have been the subject of several studies focusing on radiation effects. A numerical study of soot formation in counterflow ethylene diffusion flames at atmospheric pressure was conducted using detailed chemistry and realistic thermal and transport properties in [15]. Soot kinetics were modeled using a semi-empirical two-equation model, and radiation heat transfer was calculated using a discrete ordinates method coupled with a statistical narrow band correlated-k spectral model. The individual effects of gas and soot radiation on soot formation were isolated, and it was determined that gas radiation played a more significant role than soot radiation in this configuration. Radiation properties of the molecular gases and soot particles, including reabsorption of radiation, were considered in addition to detailed chemistry and soot models in simulations reported in [16]. The effects of radiation, including those of reabsorption, were found to be significant, particularly at lower stretch rates. Radiation modeling had a large influence on computed soot concentrations, especially for high-pressure flames. And the neglect of radiation led to significantly higher predicted levels of soot and NO, particularly at high pressures. The existence of dual extinction limits in the presence of radiative heat losses was identified in an analytical study for this configuration [17]: kinetic extinction (at low Damköhler number) and radiative extinction (at high Damköhler number). The kinetic limit was minimally affected by radiative loss, while substantial heat loss was associated with the radiative limit. Large-activation-energy asymptotic theory was used in [18] to show that soot loading significantly decreases the flammability limits, and that the minimum value of flame stretch at the radiative extinction limit is increased by more than one order of magnitude compared to a nonsooting flame. It was then argued that multidimensional sooting flames should be more susceptible to radiative extinction than the one-dimensional flame configurations that have been the focus of much of the fundamental combustion research. Flammability limits of opposed-flow H_2/CO (syngas) diffusion flames were investigated numerically in [19]. Various dilution gases (including CO_2, H_2O, and N_2) were used to alter the chemical and radiation environments in the flames. The radiation effects of H_2O and CO_2 were found to play important roles in flame extinction at low strain rates. Extinction limits of counterflow nonpremixed flames with normal- and high-temperature oxidizers were studied experimentally and numerically in [20]. The influence of radiative heat loss on stretch extinction limits of oxygen-enriched flames and air flames was investigated by comparing results from an optically thin radiation model with those from an adiabatic (neglecting radiation) model. The influence of radiative heat loss on stretch extinction limits was relatively small for the conditions that were investigated there.

A second one-dimensional configuration that has been studied extensively is a spherical diffusion flame, where a steady or transient spherical nonpremixed flame is investigated. Radiative extinction of spherical diffusion flames in microgravity was investigated experimentally and numerically in [21]. Both normal (fuel flowing into oxidizer from a porous sphere) and inverse (oxidizer flowing into fuel) flames were studied. Radiative heat loss was dominated by the combustion products downstream of the flame, and was found to scale with flame surface area rather than volume. For large transient flames the heat-release rate also scaled with surface area, so that the radiative loss fraction was largely independent of flow rate. Peak temperatures

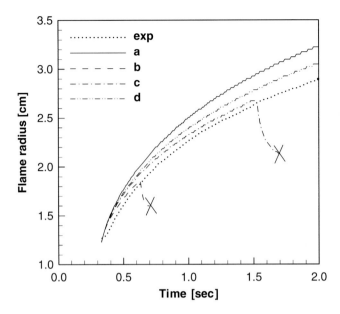

Fig. 4.3 Computed and measured flame radius versus time for a time-dependent spherical diffusion flame of C_2H_4 into diluted air (21 % O_2, 17 % N_2, 21.3 % CO_2, 40.7 % He). Simulation results are shown for four different radiation models: (*a*) adiabatic; (*b*) gray gas + optically thin model; (*c*) wide-band model; (*d*) statistical narrow band model [23]

at the onset of extinction were approximately 1100 K, which is significantly lower than for kinetic extinction. A key observation was that radiative heat losses can drive transient extinction in this configuration, and that this is not only because radiative losses increase with time, but also because the heat-release rate decreases as the flame expands away from the burner. Transient measurements of spherical diffusion flames were reported in [22]. There radiatively participating (CO_2) or nonparticipating (N_2) diluents were introduced on either the fuel side or on the oxidizer side to vary the radiation environment. When CO_2 was introduced on the oxidizer side, radiative reabsorption strengthened the flame, while CO_2 on the fuel side increased the radiative heat losses without significantly affecting reabsorption. In [23], a one-dimensional transient model of spherical diffusion flames in microgravity was developed, with variations in the spectral radiation property model and in the RTE solution method (Fig. 4.3). Based on comparisons with experiment, a statistical narrow band spectral model and a discrete ordinates method were selected for subsequent parametric studies reported in [24]. In the parametric studies, the model was exercised to explore the effects of fuel- versus oxidizer-side dilution on transient flame development and extinction. It was found that oxidizer-side dilution had a stronger influence on flame transient behavior compared to fuel-side dilution. Further parametric studies were performed using different diluents (CO_2 versus N_2 versus He) to isolate and quantify chemical versus thermal-diffusive versus radiation effects. A critical flame temperature at extinction of 1130 K was found in all cases, suggesting that extinction of spherical diffusion flames is governed by local conditions in the flame zone rather than by the volumetric radiative heat transfer. It was further found that a steady-state spherical diffusion flame cannot exist in microgravity, because the flame continues to grow at the outer edge while the flame temperature continues to drop due to radiative heat loss until the flame temperature reaches the critical value and extinguishes. The extinction of steady spherical diffusion flames stabilized by a spherical porous burner was investigated using high-activation-energy asymptotics in [25]. There an optically thin radiation model was used. Four model flames having the same adiabatic flame temperature and fuel consumption rate but different stoichiometric mixture fractions and flow directions (normal and inverse) were investigated, and extinction regimes (kinetic and radiative) were mapped out. For flames with low radiation intensity, extinction was dominated by residence time, so that the high-flow-rate flames were more susceptible to extinction, while the opposite trend was found for flames having strong radiative heat loss. Ten different normal and inverse spherical diffusion flames in microgravity were simulated in [26], and results were compared to corresponding drop-tower measurements. The flames spanned a broad range of stoichiometric mixture fraction, adiabatic flame temperature, and stoichiometric scalar dissipation rate. All flames were initially sooty (ethylene fuel), and passed through their sooting limit over the course of the experiment. Radiation was modeled using a detailed absorption/emission statistical narrow band model coupled with a discrete ordinates method. For ethylene flames with sufficiently long flow times, it was found that soot formation coincided with regions where the C/O atom ratio and temperature exceeded critical values of 0.53 and 1305 K, respectively.

Recent droplet-combustion experiments on the International Space Station revealed a previously unobserved radiation-related phenomenon in spherical diffusion flames surrounding isolated liquid droplets of n-heptane and other normal-alkane fuels [27, 28]. The experiments included three different fuels, a range of initial droplet sizes and pressures, and various diluents and oxygen levels. Under some conditions (e.g., large n-heptane droplets), following radiative extinction of the visible ("hot") flame, the flame continued to burn in a quasi-steady low-temperature regime (a "cool" flame') governed by negative-temperature-coefficient chemistry (where reactivity decreases with increasing temperature). At higher pressures, multiple cycles of this behavior were observed in some cases. In contrast, small droplets did not exhibit radiative extinction. Models were presented in both papers that reproduced the experimental trends, and that were used to explain the underlying physical processes, which required consideration of ketohydroperoxide kinetics.

4.2   Two-Dimensional Laminar Flames

As in the previous section, radiation effects in premixed flames are discussed first, followed by nonpremixed (diffusion) flames.

4.2.1   2D Laminar Premixed Flames

A standard configuration for measuring flammability limits is a vertical tube of 50 mm diameter and 1.8 m length, open at the bottom end and closed at the top. The tube is filled with a homogeneous premixed reactant mixture, and the mixture is considered to be flammable if after ignition at the bottom end, the flame propagates all the way to the top of the tube. Of particular interest is the lean flammability limit: the leanest mixture that supports flame propagation. Two-dimensional models have been developed to better understand the physical processes, including the contributions of radiative heat transfer [29, 30]. An optically thin gas and a transparent or nonreflecting tube wall were used in [30] to account (approximately) for radiation losses from CO_2 and H_2O. It was found that the effect of radiation losses decreased with the radius of the tube. Numerical results and approximate analyses showed that in the absence of radiation, a very lean flame failed to propagate after recirculation of the burned gas reached the reaction region and changed the flame structure. This condition was not realized for the standard tube diameter, but it appeared to account for the measured flammability limits in a tube of approximately half the diameter of the standard tube.

Two-dimensional unsteady simulations were performed to elucidate the effects of radiation on the dynamic behavior of cellular premixed flames at low Lewis numbers, where a disturbance was superimposed on an initially planar flame and a cellular structure developed through an intrinsic instability [31]. The average cell size of the nonadiabatic flame (with radiation) was slightly smaller than that for the adiabatic flame, even though the critical wavelength of the nonadiabatic flame was larger than that of the adiabatic flame. This suggested that the radiation had a pronounced influence on the dynamics of premixed flames at low Lewis numbers. The average burning velocity of the nonadiabatic cellular flame was smaller than that of the adiabatic cellular flame, because the local burning velocity of planar flames decreases due to radiation and the dynamic behavior of cellular flames becomes stronger.

The role of radiation heat transfer in flame propagation and transition to detonation in dust clouds was explored in [32] using a two-dimensional model that considered radiative exchange between particles using a discrete exchange model, and between the solid and the gas phases. A key mechanism for flame propagation was found to be the radiant heating to ignition of the dust cloud contained in the preheat zone, which could eventually lead to an explosion (deflagration to detonation transition).

4.2.2 2D Laminar Diffusion Flames

The influence of high-intensity radiation on the distribution of soot in two-dimensional nonpremixed, partially premixed, and fully premixed ethylene/air laminar flames was investigated experimentally in [33]. Radiation at flux levels comparable to those achievable with concentrated solar radiation was provided by a CO_2 laser at a wavelength that allowed isolation of three possible mechanisms of the influence on soot production: molecular excitation of the fuel, irradiation of the soot, and irradiation of soot precursors. The high-energy radiation increased the peak volume fraction of the soot by up to 250 %, with the dominant mechanism being identified as heating of the fuel.

The structure and extinction of counterflow diffusion flames were investigated through comparisons of experiments and calculations for normal gravity and microgravity conditions in [34]. The minimum nitrogen volume fraction in the fuel stream needed to ensure suppression (radiative extinction) for all strain rates in microgravity was higher than the corresponding value in normal gravity. The simulations yielded insight into the differences between microgravity and normal gravity suppression results, which were attributed to buoyancy effects, and two-dimensional simulations revealed the inadequacy of a one-dimensional model to explain the microgravity suppression results.

An axisymmetric coflow laminar diffusion flame where fuel issues from a central jet into a coflowing oxidizer stream (the "normal" configuration), or where the fuel and oxidizer streams are switched (fuel on the outside—an "inverse" flame), is a commonly investigated configuration for studying soot and radiation processes in flames [35–43]. In an early experimental study [35], the transition between a "smoking" luminous flame (where soot breaks through and is emitted downstream) and a "nonsmoking" luminous flame (where all soot is oxidized in the soot burnout region, and essentially no soot is emitted from the flame) was investigated by making measurements close to the smoke point for a range of fuels. It was observed that smoke was emitted when the temperature in the soot burnout region dropped below approximately 1300 K. A smoking flame close to the smoking limit was made to stop emitting smoke when the temperature in the burnout region was raised slightly locally by absorption of energy from a CO_2 laser, thus demonstrating strong coupling between radiative heat transfer and soot for flames close to the smoke threshold. Numerical calculations of an axisymmetric coflow laminar methane/air diffusion flame at atmospheric pressure were conducted using detailed transport properties and gas-phase chemistry, a semi-empirical two-equation soot model, and nongray radiative heat transfer by CO_2, H_2O, CO, and soot using a discrete ordinates method in [36]. Results from five different radiative property models were compared to assess accuracy versus computational efficiency tradeoffs. Global effects of radiation were relatively small (an optically thin model yielded temperatures approximately 17 K lower and integrated soot levels approximately 6 % lower than the best radiation model), and reabsorption had only a minor influence on computed soot volume fractions. The interaction between soot and

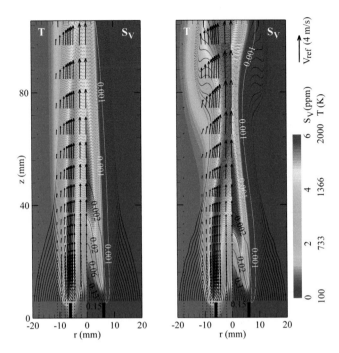

Fig. 4.4 Effect of soot radiation on the dynamics of a 15 % acetylene flame. *Left*: Steady-state flame computed when soot radiation was considered. *Right*: Instantaneous snapshot of unsteady flame structure computed when soot radiation was ignored. Velocity vectors are superimposed on the temperature field on the *left half* of each image, and isocontours of acetylene (*black*) and OH (*white*) concentrations are superimposed on the soot field on the *right half* of each image. Instantaneous particle locations are also superimposed to show streaklines [39]

NO formation in a laminar axisymmetric coflow ethylene/air diffusion flame was investigated numerically in [38]. The results showed that the formation of NO had little influence on soot formation, while increased soot significantly suppressed NO formation. The influence of soot on NO formation resulted from both a radiation-induced thermal effect (higher soot corresponding to higher radiative emission), and from a reaction-induced chemical effect (competition for C_2H_2 between soot and NO). In [39] it was demonstrated using unsteady two-dimensional numerical simulations, and subsequently confirmed experimentally, that soot radiation could influence oscillations ("flicker") resulting from a buoyancy-induced instability in coflow laminar diffusion flames, to the point that the oscillation could be completely suppressed for sufficiently high soot levels (Fig. 4.4). The effects of radiation heat transfer and radiative reabsorption on flame structure and soot formation in a coflow laminar ethylene/air diffusion flame at normal and microgravity were investigated numerically in [40]. The simulations featured detailed chemistry and transport properties, an acetylene-based soot model, and a statistical narrow band correlated-k nongray gas radiation model. Radiation heat transfer and reabsorption in the microgravity flame were found to be much more important than in the normal-gravity flame.

A comprehensive simulation study focusing on radiation effects in axisymmetric coflow laminar diffusion flames was reported in [42]. Twenty-four different flames were simulated, including different fuels (C1-to-C3 hydrocarbons), normal (NDF) and inverse (IDF) diffusion flames, normal gravity (1 g) and microgravity (0 g), and covering a wide range of conditions with respect to radiative heat transfer. Key elements of the simulations included a steady laminar flamelet model, a semi-empirical two-equation acetylene–benzene-based soot model, and the statistical narrow band correlated-k spectral model with a finite-volume method for thermal radiation. For all flames, consideration of thermal radiation was reported to be crucial in providing accurate predictions of temperatures and soot levels compared to experiment. Radiation effects were relatively more prominent in NDFs compared to IDFs, and in microgravity compared to standard gravity. Molecular gas radiation dominated in the weakly sooting IDFs and in the methane and ethane NDFs, whereas soot radiation dominated in the other flames. However, neither contribution was negligible in any of the flames, with the exception of the 1 g IDFs, where soot radiation could be ignored. The optically thin approximation was found to be satisfactory for cases where the optical thickness (based on flame radius and Planck-mean absorption coefficient) was less than 0.05; this included the IDFs and most of the 1 g NDFs, but not the 0 g NDFs. For the 0 g NDFs, a gray approximation was sufficient for soot, but not for the combustion gases. In that case, either the nongray- or the gray-soot versions of the full-spectrum correlated-k spectral model could be substituted for the narrow band version with a significant ($\sim$20$\times$) reduction in CPU time. Examples of results for 0 g flames are shown in Fig. 4.5.

References

1. Y. Ju, G. Masuya, F. Liu, Y. Hattori, D. Riechelmann, Asymptotic analysis of radiation extinction of stretched premixed flames. Int. J. Heat Mass Transf. **43**, 231–239 (2000)
2. Y. Ju, G. Masuya, P.D. Ronney, Effects of radiative emission and absorption on the propagation and extinction of premixed gas flames. Proc. Combust. Inst. **27**, 2619–2626 (1998)
3. A. Haghiri, M. Bidabadi, Dynamic behavior of particles across flame propagation through micro-iron dust cloud with thermal radiation effect. Fuel **90**, 2413–2421 (2011)
4. Z. Chen, Effects of radiation and compression on propagating spherical flames of methane/air mixtures near the lean flammability limit. Combust. Flame **157**, 2267–2276 (2010)
5. J. Santner, F.M. Haas, Y. Ju, F.L. Dryer, Uncertainties in interpretation of high pressure spherical flame propagation rates due to thermal radiation. Combust. Flame **161**, 147–153 (2014)
6. J. Jayachandran, R. Zhao, F.N. Egolfopoulos, Determination of laminar flame speed using stagnation and spherically expanding flames: molecular transport and radiation effects. Combust. Flame **161**, 2305–2316 (2014)
7. H. Yu, W. Han, J. Santner, X. Gou, C.H. Sohn, Y. Ju, Z. Chen, Radiation-induced uncertainty in laminar flame speed measured from propagating spherical flames. Combust. Flame **161**, 2815–2824 (2014)

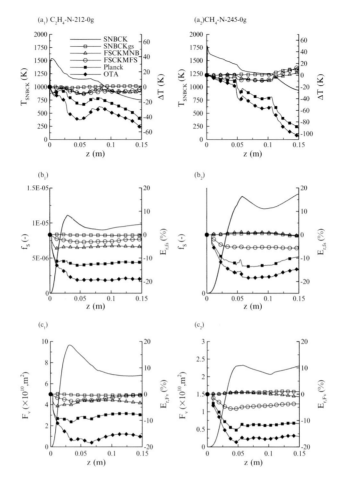

Fig. 4.5 Influences of radiative property models on computed temperatures and soot levels for C_2H_4 (*left*) and CH_4 (*right*) normal coflow laminar diffusion flames in microgravity. (**a**) Temperature profile from the baseline SNBCK spectral model (*line without symbols*), and temperature difference profiles with respect to the baseline for other spectral models (*lines with symbols*). (**b**) Soot volume fraction profile from the baseline SNBCK spectral model, and soot volume fraction percent difference profiles with respect to the baseline for other spectral models. (**c**) Integrated soot volume fraction profile from the baseline SNBCK spectral model, and integrated soot volume fraction percent difference profiles with respect to the baseline for other spectral models. For (**a**) and (**b**), z is a coordinate along the pathline corresponding to the maximum soot volume fraction. For (**c**), z is the axial distance from the burner [42]

8. L. Qiao, Transient flame propagation process and flame-speed oscillation phenomenon in a carbon dust cloud. Combust. Flame **159**, 673–685 (2012)
9. M. Bidabadi, A.V. Azad, Effects of radiation on propagating spherical flames of dust-air mixtures. Powder Technol. **276**, 45–59 (2015)
10. Z. Chen, Y. Ju, Combined effects of curvature, radiation, and stretch on the extinction of premixed tubular flames. Int. J. Heat Mass Transf. **51**, 6118–6125 (2008)

11. M.A.A. Mendes, J.M.C. Pereira, J.C.F. Pereria, A numerical study of the stability of one-dimensional laminar premixed flames in inert porous media. Combust. Flame **153**, 525–539 (2008)
12. I. Schoegl, Radiation effects on flame stabilization on flat flame burners. Combust. Flame **159**, 2817–2828 (2012)
13. J.S. T'ien, Diffusion flame extinction at small stretch rates: the mechanism of radiative loss. Combust. Flame **65**, 31–34 (1986)
14. H.Y. Wang, C.K. Law, On intrinsic oscillation in radiation-affected diffusion flames. Proc. Combust. Inst. **31**, 979–987 (2007)
15. F. Liu, H. Guo, G.J. Smallwood, M. El Hafi, Effects of gas and soot radiation on soot formation in counterflow ethylene diffusion flames. J. Quant. Spectrosc. Radiat. Transf. **84**, 501–511 (2004)
16. X.L. Zhu, J.P. Gore, Radiation effects on combustion and pollutant emissions of high-pressure opposed flow methan/air diffusion flames. Combust. Flame **141**, 118–130 (2005)
17. H.Y. Wang, W.H. Chen, C.K. Law, Extinction of counterflow diffusion flames with radiative heat loss and nonunity Lewis numbers. Combust. Flame **148**, 100–116 (2007)
18. P. Narayanan, H.R. Baum, A. Trouvé, Effect of soot addition on extinction limits of luminous laminar counterflow diffusion flames. Proc. Combust. Inst. **33**, 2539–2546 (2011)
19. H.-Y. Shih, J.-R. Hsu, Y.-H. Lin, Computed flammability limits of opposed-jet H_2/CO syngas diffusion flames. Int. J. Hydrog. Energy **39**, 3459–3468 (2014)
20. X. Li, L. Jia, T. Onishi, P. Grajetzki, H. Nakamura, T. Tezuka, S. Hasegawa, K. Maruta, Study on stretch extinction limits of CH_4/CO_2 versus high temperature O_2/CO_2 counterflow non-premixed flames. Combust. Flame **161**, 1526–1536 (2014)
21. K.J. Santa, B.H. Chao, P.B. Sunderland, D.L. Urban, D.P. Stocker, R.L. Axelbaum, Radiative extinction of gaseous spherical diffusion flames in microgravity. Combust. Flame **151**, 665–675 (2007)
22. M.K. Chernovsky, A. Atreya, H.G. Im, Effect of CO_2 diluent on fuel versus oxidizer side of spherical diffusion flames in microgravity. Proc. Combust. Inst. **31**, 1005–1013 (2007)
23. S. Tang, M.K. Chernovsky, H.G. Im, A. Atreya, A computational study of spherical diffusion flames in microgravity with gas radiation. Part I: model development and validation. Combust. Flame **157**, 118–126 (2010)
24. S. Tang, H.G. Im, A. Atreya, A computational study of spherical diffusion flames in microgravity with gas radiation. Part II: parametric studies of the diluent effects on flame extinction. Combust. Flame **157**, 127–136 (2010)
25. Q. Wang, B.H. Chao, Kinetic and radiative extinctions of spherical burner-stabilized diffusion flames. Combust. Flame **158**, 1532–1541 (2011)
26. V.R. Lecoustre, P.B. Sunderland, B.H. Chao, R.L. Axelbaum, Numerical investigation of spherical diffusion flames at their sooting limits. Combust. Flame **159**, 194–199 (2012)
27. V. Nayagam, D.L. Dietrich, M.C. Hicks, F.A. Williams, Cool-flame extinction during n-alkane droplet combustion in microgravity. Combust. Flame **162**, 2140–2147 (2015)
28. T.I. Farouk, M.C. Hicks, F.L. Dryer, Multistage oscillatory "cool flame" behavior for isolated alkane droplet combustion in elevated pressure microgravity combustion. Proc. Combust. Inst. **35**, 1701–1708 (2015)
29. Y. Shoshin, J. Jarosinski, On extinction mechanism of lean limit methane-air flame in a standard flammability tube. Proc. Combust. Inst. **32**, 1043–1050 (2009)
30. F.J. Higuera, V. Muntean, Effect of radiation losses on very lean methane/air flames propagating upward in a vertical tube. Combust. Flame **161**, 2340–2347 (2014)
31. S. Kadowaki, H. Takahashi, H. Kobayashi, The effects of radiation on the dynamic behavior of cellular premixed flames generated by intrinsic instability. Proc. Combust. Inst. **33**, 1153–1162 (2011)
32. R.B. Moussa, M. Guessasma, C. Proust, K. Saleh, J. Fortin, Thermal radiation contribution to metal dust explosions. Procedia Eng. **102**, 714–721 (2015)
33. P.R. Medwell, G.J. Nathan, Q.N. Chan, Z.T. Alwahabi, B.B. Dally, The influence on the soot distribution within a laminar flame of radiation at fluxes of relevance to concentrated solar radiation. Combust. Flame **158**, 1814–1821 (2011)

34. A. Hamins, M. Bundy, C.B. Oh, S.C. Kim, Effect of buoyancy on the radiative extinction limit of low-strain-rate nonpremixed methane-air flames. Combust. Flame **151**, 225–234 (2007)
35. J.H. Kent, H.G. Wagner, Why do diffusion flames emit smoke? Combust. Sci. Technol. **41**, 245–269 (1984)
36. F. Liu, H. Guo, G.J. Smallwood, Effects of radiation model on the modeling of a laminar coflow methane/air diffusion flame. Combust. Flame **138**, 136–154 (2004)
37. H. Guo, F. Liu, G.J. Smallwood, O.L. Gülder, Numerical study on the influence of hydrogen addition on soot formation in a laminar ethylene-air diffusion flame. Combust. Flame **145**, 324–338 (2006)
38. H. Guo, G.J. Smallwood, The interaction between soot and NO formation in a laminar axisymmetric coflow ethylene/air diffusion flame. Combust. Flame **149**, 224–233 (2007)
39. V.R. Katta, W.M. Roquemore, A. Menon, S.-Y. Lee, R.J. Santoro, T.A. Litzinger, Impact of soot on flame flicker. Proc. Combust. Inst. **32**, 1343–1350 (2009)
40. F. Liu, G.J. Smallwood, W. Kong, The importance of thermal radiation transfer in laminar diffusion flames at normal and microgravity. J. Quant. Spectrosc. Radiat. Transf. **112**, 1241–1249 (2011)
41. A. Fuentes, R. Henríqez, F. Nmira, F. Liu, J.-L. Consalvi, Experimental and numerical study of the effects of the oxygen index on the radiation characteristics of laminar coflow diffusion flames. Combust. Flame **160**, 786–795 (2013)
42. R. Demarco, F. Nmira, J.L. Consalvi, Influence of thermal radiation on soot production in laminar axisymmetric diffusion flames. J. Quant. Spectrosc. Radiat. Transf. **120**, 52–69 (2013)
43. D. Zhang, J. Fang, J.-F. Guan, J.-W. Wang, Y. Zeng, J.-J. Wang, Y.-M. Zhang, Laminar jet methane/air diffusion flame shapes and radiation of low air velocity coflow in microgravity. Fuel **130**, 25–33 (2014)

Chapter 5
DNS and LES of Turbulence–Radiation Interactions in Canonical Systems

A number of DNS and LES studies have focused on radiative heat transfer and turbulence–radiation interactions in canonical (highly idealized, and usually not physically realizable) turbulent reacting or nonreacting systems. In contrast to an actual combustion system or device, this allows direct manipulation and isolation of the physical processes of interest, and removes ambiguities in the interpretation of the results. While one must exercise caution in extrapolating the results from these highly simplified problems to practical turbulent flames, such investigations are useful for physics discovery and for providing data that can be used to develop and/or calibrate models that are subsequently applied to real turbulent flames and combustion devices. In this chapter, examples of DNS and LES studies of radiation and TRI in canonical configurations are presented and discussed. This includes some results for nonreacting systems, and studies with both uncoupled radiation (no feedback of the radiative source term into the CFD simulation) and with coupled radiation. Most experimental and modeling studies of TRI have focused on how turbulence affects radiative heat transfer, but the investigations with coupled radiation also allow one to explore at a fundamental level how radiation and TRI modify the structure of the turbulent flow field.

5.1 Decoupled Analyses: Post-Processed Chemistry and Radiation

TRI were investigated using DNS for an idealized nonpremixed system in [1]. The configuration was statistically homogeneous isotropic decaying turbulence in a cube with periodic boundary conditions, including a passive scalar mixture fraction field for scalar mixing (no combustion). Species compositions and temperature were mapped to the mixture fraction using a fast-chemistry assumption, so that the

© The Author(s) 2016
M.F. Modest, D.C. Haworth, *Radiative Heat Transfer in Turbulent Combustion Systems*, SpringerBriefs in Applied Sciences and Technology, DOI 10.1007/978-3-319-27291-7_5

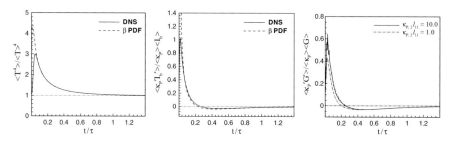

Fig. 5.1 Evolution of normalized TRI correlations with normalized time for a statistically homogeneous nonpremixed system. *Left*: Temperature self-correlation (emission TRI). *Center*: Absorption coefficient–Planck function correlation (emission TRI). *Right*: Absorption coefficient–intensity correlation (absorption TRI). Emission TRI quantities are independent of system optical thickness for this configuration, and results from a presumed beta-PDF model are shown. Absorption TRI profiles are shown for two different system optical thicknesses [1]

implied radiatively participating species concentrations and temperatures were at levels representative of those in a turbulent nonpremixed flame. DNS snapshots at discrete instants in time were then post-processed to calculate radiative heat transfer based on the mapped species concentrations and temperature, with no feedback of the radiative source term to the DNS. The radiation properties corresponded to a nonscattering fictitious gray gas with a participating-species- and temperature-dependent Planck-mean absorption coefficient that mimicked that of typical hydrocarbon–air combustion products, including a free parameter to allow the optical thickness of the system to be varied. A photon Monte Carlo method was used to solve the RTE. Individual contributions of emission and absorption TRI were isolated and quantified, focusing on the temperature self-correlation, the absorption coefficient–Planck function correlation, and the absorption coefficient–intensity correlation. For intermediate-to-large values of the optical thickness, contributions from all three correlations were found to be significant. All three terms were positive initially, but the absorption coefficient–Planck function correlation and the absorption coefficient–intensity correlation were found to be negative over a period of time as the system evolved (Fig. 5.1). Emission TRI were also estimated analytically using a beta-PDF model, where the mixture fraction PDF was presumed to follow a beta distribution (a two-parameter distribution requiring specification of the mean and the variance), and the mixture fraction mean and variance as functions of time were extracted from the DNS; the model results were in good agreement with the DNS data after a short initial transient (Fig. 5.1). This suggested that a relatively simple model for emission TRI in nonpremixed systems might be developed based on modeled equations for mixture fraction mean and variance for high-Damköhler-number (fast chemistry compared to turbulence) systems.

In a series of papers [2–5], DNS databases for nonreacting turbulent flows were analyzed to guide the development of subfilter-scale models of TRI for large-eddy simulation. In [2–4], the configuration was a homogeneous isotropic nonpremixed system with a passive scalar, similar to the one investigated in [1], but with forcing to maintain a statistically stationary (versus decaying) system. The DNS velocity and passive scalar fields were mapped/rescaled to provide velocity, temperature, and participating-gas (CO_2) fields having statistics that were representative of those downstream of the main combustion zone in a nonpremixed turbulent jet flame. A ray-tracing method was then used to solve the RTE along lines of sight through the mapped/rescaled snapshots of the DNS fields, with a correlated-k-distribution model for spectral properties (uncoupled radiation), and various statistics were compiled. Radiation fields computed using the DNS velocity, species, and temperature fields directly (where all turbulent fluctuations are fully resolved) were compared with radiation fields computed using spatially filtered values of the DNS velocity, species, and temperature fields (corresponding to what would be done in LES) to isolate and quantify various TRI contributions. In [3], results were presented for different levels of rms turbulence velocity, different levels of rms species and temperature fluctuations, and different spatial filter sizes. Further analysis was reported in [4], including moments of the radiation intensity, Planck-mean and incident-mean absorption coefficients, and emission and absorption TRI correlations. In [2], additional parametric studies were performed with variations in the mean temperature level, and results from an LES model that neglected the influences of subfilter-scale fluctuations, in addition to the DNS and filtered DNS results (Fig. 5.2). For the range of conditions simulated in these studies, contributions from subfilter-scale fluctuations to emission TRI were found to be significant (implying that a model for subfilter-scale emission TRI would be required in LES), while contributions from subfilter-scale fluctuations to absorption TRI were negligible (implying that the OTFA would be sufficient). For some conditions, the subfilter-scale temperature self-correlation and absorption coefficient–temperature correlation had opposite,

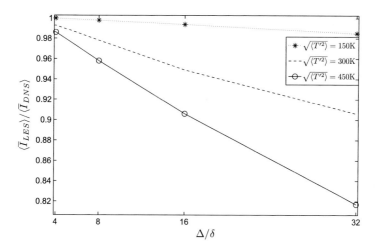

Fig. 5.2 Influence of the temperature variance and filter size on the ratio of the mean radiation intensity from LES without a subfilter-scale TRI model to the mean radiation intensity from filtered DNS, for a fixed mean temperature of 1500 K [2]

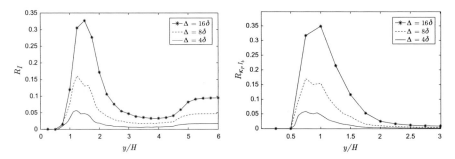

Fig. 5.3 Profiles of ratios of subfilter-scale TRI contributions to resolved-scale TRI contributions in a planar turbulent jet for different filter sizes [5]. *Left*: Radiative intensity ratio. *Right*: Radiative emission ratio

and largely offsetting, effects, so that neglecting TRI altogether yielded satisfactory results for sufficiently low rms temperature fluctuation levels and sufficiently fine meshes (filter sizes).

A similar post-processing analysis approach was applied to passive scalar nonreacting DNS data in a different configuration (a temporally evolving planar turbulent jet) in [5], including variations in overall system optical thickness (Fig. 5.3). There it was found that the subfilter-scale TRI were relatively more important near the edge of the jet (locally high mean temperature gradients), and relatively less important close to the jet centerline.

Radiative heat transfer in an idealized fully developed turbulent channel flow between two infinite, parallel, stationary plates at different fixed temperatures was studied using LES in [6]. There the same gray spectral model was used as in [1], the RTE was solved using a P1 method, and the computed radiative source term was fed back into the energy equation (thereby influencing the computed temperature field), but the temperature was passive with respect to the fluid mechanics (incompressible hydrodynamics). The passive scalar LES was validated against DNS data from the literature (mean and rms velocity and scalar profiles) for the same configuration. Then radiation and TRI results for two cases were presented. In the first case ("nonreacting"), the composition was spatially homogeneous while the temperature varied due to heat transfer, and a parametric study was performed with variations in the system global optical thickness. For the nonreacting case, the computed mean and rms temperature profiles were affected significantly by radiation, but TRI (even the contributions of the resolved-scale fluctuations) were negligible. In the second case ("reacting"), the composition was nonuniform and the boundary conditions for a passive scalar were specified to correspond to a mixing layer between "fuel" and "oxidizer." A fast-chemistry assumption was invoked to map the mixture fraction to radiatively participating species compositions and temperature, and resolved-scale TRI-related quantities were extracted for the reacting mixing layer with variations in the global system optical thickness. For the reacting case, composition and temperature fluctuations were amplified compared to the nonreacting case. There strong emission TRI were seen for all optical thicknesses,

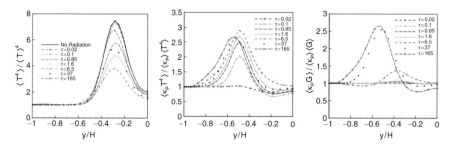

Fig. 5.4 Profiles of normalized resolved-scale TRI quantities in LES of a reacting turbulent mixing layer for different system optical thicknesses [6]. *Left*: Temperature self-correlation (emission TRI). *Center*: Absorption coefficient–Planck function correlation (emission TRI). *Right*: Absorption coefficient–intensity correlation (absorption TRI)

nonmonotonic behavior for emission TRI was observed with variations in optical thickness, and absorption TRI were significant only for optically thick systems (Fig. 5.4). A similar planar channel flow configuration was investigated using DNS in [7], including comparisons of results for two different Reynolds numbers. The principal metric was the ratio of the mean radiative intensity to the radiative intensity based on the mean temperature (the "enhancement ratio"). Computed enhancement ratios were between 1.0 and 1.4 for all conditions studied, with peak values being closer to the wall for the higher Reynolds number case.

5.2   Nonreacting Turbulent Flows with Coupled Radiation

DNS of nonreacting idealized fully developed turbulent channel flow between two infinite, parallel, stationary plates at different fixed temperatures with coupled radiative heat transfer was considered in [8, 9]. In [8], results from multiple simulations were presented with variations in Prandtl number (0.1, 0.71, and 2.0), Reynolds number (bulk Reynolds numbers from 2800 to 11,750), hot-wall temperature (1150 or 2050 K), pressure (1 or 40 atm), cold- and hot-wall emissivities (0.1–1.0), and with versus without radiation. A low-Mach-number compressible ideal-gas formulation was used, where temperature changes induced by radiative heat transfer influence the fluid density and other properties, and a reciprocal Monte Carlo method was used to solve the RTE. Opaque constant-emissivity walls were considered, and participating-medium radiative properties corresponded to idealized combustion products (fixed proportions of CO_2, H_2O, and N_2) with a correlated-k spectral model for 1 atm cases and the weak absorption limit for high-pressure cases. Results were analyzed to isolate gas–gas and gas–wall radiation interactions. One finding was that the classical log-law for temperature in a fully developed turbulent boundary layer was strongly modified by radiation (Fig. 5.5). Results from three of the high-pressure cases reported in [8] were analyzed further

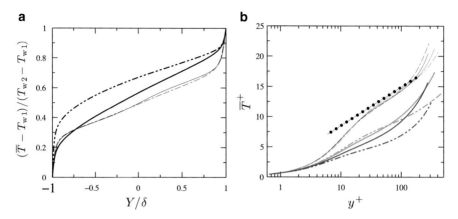

Fig. 5.5 Normalized mean temperature profiles in standard (*left*—δ is the channel half-width) and wall-scaled coordinates (*right*—the *line with symbols* corresponds to a standard wall function) for fully developed turbulent channel flow between constant-temperature plates at 40 atm [8]. The cold-wall temperature is 950 K in all cases. *Solid lines* are for a hot-wall temperature of 1150 K, and *dashed lines* are for a hot-wall temperature of 2050 K. *Bold lines* are for cases that include radiation (wall emissivity 0.8), and *thin lines* are for cases that neglect radiation. In the *right-hand plot*, *cyan color* corresponds to the cold-wall profiles and *red color* to the hot-wall profiles

in [9]; the high-pressure (40 atm) cases were selected because radiation effects were more prominent compared to the 1 atm cases. The boundary-layer structure was modified significantly, and temperature fluctuations and the turbulent heat flux were reduced with consideration of radiation. Budgets of contributions of individual terms in the transport equations for enthalpy variance and turbulent heat flux were presented to understand the observed behavior. In the absence of radiation, profiles for all Reynolds numbers collapse to a single universal profile when scaled using classical turbulence wall variables. The same scaling does not collapse the profiles when radiation is considered, and a radiation-based scaling was proposed that collapsed the profiles for cases with radiation. A turbulent-Prandtl-number-based model was proposed and tested to account for the strong influence of radiation on turbulent wall heat transfer. In the absence of radiation and with a (molecular) Prandtl number of $Pr = 0.71$ (corresponding to air), the value of the turbulent Prandtl number Pr_t in fully developed channel flow is approximately unity close to the wall, and decreases in the wall-normal direction to an asymptotic value of approximately 0.7 for sufficiently high Reynolds numbers. The behavior of Pr_t with consideration of radiation is quite different. There the value of Pr_t is greater than unity at the wall, decreases with increasing distance from the wall up to a point, then may or may not increase again, depending on the relative importance of convection versus radiation. Results from the proposed model for Pr_t are compared with the DNS results for several cases in Fig. 5.6.

Radiation and TRI in a fully developed supersonic channel flow (Mach numbers up to ~2.9) were explored in [10]. LES was performed based on the filtered compressible Navier–Stokes equations for a nonreacting ideal-gas medium (pure

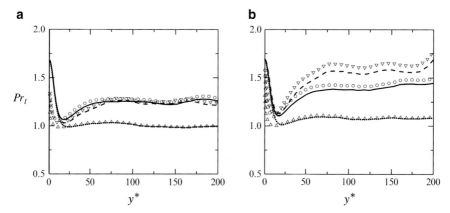

Fig. 5.6 Profiles of turbulent Prandtl number on the cold side (**a**) and the hot side (**b**) for fully developed turbulent channel flow between constant-temperature plates at 40 atm, with variations in bulk Reynolds number Re_b and hot-wall temperature T_h [9]. *Symbols* are DNS results, and *lines* are model results. The cold-wall temperature is 950 K in all cases. *Circles* and *solid lines*: $Re_b = 5850$, $T_h = 1150$ K. *Downward triangles* and *dashed lines*: $Re_b = 5850$, $T_h = 2050$ K. *Upward triangles* and *dotted lines*: $Re_b = 11,750$, $T_h = 1150$ K

water vapor), including the radiative source term in the energy equation. Spectral properties for an absorbing/emitting ideal-gas medium were considered using a statistical narrow band correlated-k model, and some results from a gray model were also presented for comparison. Walls were treated as black, constant-temperature (1000 K) surfaces. The RTE was solved using a discrete ordinates method, and subfilter-scale TRI were neglected. Profiles of mean quantities, Reynolds stresses, and budgets of various quantities including energy showed that radiation effects counteracted compressibility effects in some cases. Radiation effects were relatively small overall for the conditions that were considered, although differences could be seen in the computed mean and rms temperature profiles with versus without radiation. This work was extended to compare TRI effects in a fully developed supersonic channel flow and in a fully developed supersonic axisymmetric pipe flow (both at Mach numbers of approximately 1.2) in [11]. There the medium was considered to be gray, with a temperature-dependent Planck-mean absorption coefficient having the same functional form as that used in [6], and the value of the free parameter was varied to correspond to absorption coefficient values that were up to 10 times the physical value for water vapor, to amplify radiation effects. Results for two different optical thicknesses were presented, and comparisons between channel flow and pipe flow (including normalized mean velocity, Reynolds stress, mean temperature, and energy budget terms profiles) were made for the same friction Reynolds numbers and Mach numbers. Large differences were found between the two configurations, and physical explanations for the differences were offered.

5.3   Turbulent Reacting Flows with Coupled Radiation

DNS for a statistically one-dimensional propagating turbulent premixed flame including coupled radiation heat transfer was reported in [12, 13]. A reacting ideal-gas mixture was considered, with single-step irreversible finite-rate chemistry, a gray participating medium with the same composition- and temperature-dependent Planck-mean absorption coefficient as that used in [6], and a photon Monte Carlo method to solve the RTE. In [12], two-dimensional simulations were performed for a no-radiation case and for cases with radiation for three different optical thicknesses (optically thin, intermediate, and thick), and a three-dimensional optically intermediate case was included for comparison purposes. Profiles of temperature self-correlation, absorption coefficient–Planck function correlation, and absorption coefficient–intensity correlation were presented and discussed, and results were compared with those from a simple flame-sheet-based model. In this configuration, the temperature self-correlation contribution (emission TRI) was primarily a function of the ratio of burned-gas temperature to unburned-gas temperature, and was the dominant contribution to TRI only in the optically thin limit. Even in the most optically thin case considered, the absorption coefficient–Planck function correlation and absorption coefficient–intensity correlation were not negligible. At intermediate values of optical thickness, contributions from all three correlations were found to be important (Fig. 5.7). The same configuration was considered in [13]. There radiation-induced modifications to the turbulent flame structure were observed, but the main focus was the development and validation of a high-order photon Monte Carlo method for DNS.

The same numerical methods, thermochemical models, and radiation models were used in DNS of statistically one-dimensional turbulent nonpremixed flames with coupled radiation in [14]. The temperature self-correlation, the absorption coefficient–Planck function correlation, and the absorption coefficient–intensity correlation were examined for intermediate-to-large values of the optical thickness.

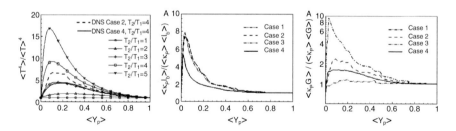

Fig. 5.7 Profiles of normalized TRI quantities as functions of mean progress variable for a statistically one-dimensional premixed turbulent flame [12]. Case 1: 2D, optically thin; Case 2: 2D, optically intermediate; Case 3: 2D, optically thick; Case 4: 3D, optically intermediate. *Left*: Temperature self-correlation for Cases 2 (*dashed line*) and 4 (*solid line*), and profiles from a flame-sheet approximation for several values of the temperature ratio (*lines with symbols*). *Center*: Absorption coefficient–Planck function correlation. *Right*: Absorption coefficient–intensity correlation

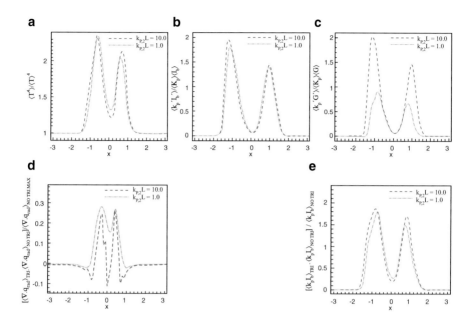

Fig. 5.8 Profiles of normalized TRI quantities at one instant in time from DNS for a statistically one-dimensional turbulent nonpremixed flame, for an optically intermediate case and an optically thick case [14]. (**a**) Temperature self-correlation. (**b**) Absorption coefficient–Planck function correlation. (**c**) Absorption coefficient–intensity correlation. (**d**) Net overall TRI. (**e**) Net emission TRI

Contributions from all three correlations were found to be not negligible, but the relative importance of their contributions varied with optical thickness. Figure 5.8 shows profiles of various TRI quantities at one instant in time. Ideally, mean quantities should be symmetric about $x = 0$; there are departures from symmetry because of the limited sample size in the DNS. In [15], the same one-dimensional nonpremixed system was studied using a third-order spherical harmonics method (P3 approximation) to solve the RTE. The focus there was on the development and validation of the P3 method for DNS, and the TRI results were similar to those reported in [14] using the Monte Carlo method.

LES of nonreacting and reacting planar mixing layers with coupled radiation was reported in [16, 17]. Simulations assuming ideal-gas mixtures and gray-gas radiation properties with artificially increased Planck-mean absorption coefficients (following [6]) were performed to enhance the radiation effects, and a discrete ordinates method was used to solve the RTE, neglecting the influences of subfilter-scale fluctuations on radiation. For the nonreacting case, the low-speed stream consisted of nitrogen at 1000 K and the high-speed stream was pure water vapor at 2000 K. For the reacting case, the low-speed stream was a mixture of hydrogen and nitrogen and the high-speed stream was a mixture of oxygen and nitrogen, with both streams premixed such that the free-stream densities were the same and the stoichiometric mixture fraction value was 0.3. An equilibrium assumption was used to model the chemistry

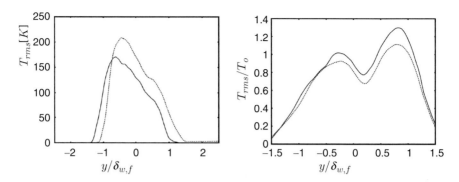

Fig. 5.9 Profiles of rms temperature fluctuations with (*dashed lines*) and without (*solid lines*) radiation from LES of nonreacting and reacting mixing layers [16]. *Left*: Nonreacting case (not normalized). *Right*: Reacting case (normalized)

in terms of the mixture fraction. In both the nonreacting and the reacting cases, radiative heat transfer was found to affect the fluctuation levels of thermodynamic variables, Reynolds stresses, and Reynolds-stress budget terms. Budgets of source terms in the transport equation for the temperature variance were analyzed to explain the observed decrease of temperature variance in the reacting case and its increase in the nonreacting case (Fig. 5.9).

5.4   Summary

A few general conclusions regarding TRI can be drawn from the results for canonical systems that have been reviewed in this chapter. While the extent to which these conclusions can be extrapolated to practical turbulent flames is unclear, they nonetheless provide some initial guidance on what one should look for and expect in more realistic turbulent reacting flows.

For low-speed nonreacting systems, radiation effects can be strong, but influences of unresolved turbulent fluctuations in RANS or in LES are usually negligible, so that TRI modeling is not required. This is because the fluctuation levels for participating medium concentrations and temperature are relatively small in low-Mach-number nonreacting turbulent flows. TRI are evident in atmospheric-scale inert turbulent flows, where the optical path lengths are orders of magnitude longer compared to combustion systems [18]. The fluctuation levels are higher in high-speed and/or reacting flows, and TRI are commensurately more prominent there. Net emission can increase significantly with consideration of TRI in both RANS and in LES, and the dominant contribution in most cases (for gray radiation, at least) is from temperature fluctuations (temperature self-correlation). However, the absorption coefficient–Planck function contribution to emission TRI also is not negligible in most cases, while absorption TRI are important only in optically very thick systems. While the temperature self-correlation is always positive, the absorption coefficient–Planck function correlation and the absorption coefficient–intensity correlation can be either positive or negative, so that (for example) the absorption coefficient–Planck function contribution can partially offset the temperature self-correlation emission enhancement. Temperature fluctuations are usually suppressed by TRI (an exception being the increase noted for a nonreacting supersonic mixing layer in Fig. 5.9), and there can be strong interactions between radiative and convective heat transfer in turbulent boundary layers resulting from radiation-induced modifications to the turbulent flow structure.

References

1. K.V. Deshmukh, D.C. Haworth, M.F. Modest, Direct numerical simulation of turbulence–radiation interactions in a statistically homogeneous nonpremixed combustion system. Proc. Combust. Inst. **31**, 1641–1648 (2007)
2. M. Roger, P.J. Coelho, C.B. da Silva, The influence of the non-resolved scales of thermal radiation in large eddy simulation of turbulent flows: a fundamental study. Int. J. Heat Mass Transf. **53**, 2897–2907 (2010)
3. M. Roger, C.B.D. Silva, P.J. Coelho, Analysis of the turbulence–radiation interactions for large eddy simulations of turbulent flows. Int. J. Heat Mass Transf. **52**, 2243–2254 (2009)
4. C.B. da Silva, I. Malico, P.J. Coelho, Radiation statistics in homogeneous isotropic turbulence. New J. Phys. **11**, 093001–1–34 (2009)
5. M. Roger, P.J. Coelho, C.B. da Silva, Relevance of the subgrid-scales for large eddy simulations of turbulence-radiation interactions in a turbulent plane jet. J. Quant. Spectrosc. Radiat. Transf. **112**, 1250–1256 (2011)
6. A. Gupta, M.F. Modest, D.C. Haworth, Large-eddy simulation of turbulence–radiation interactions in a turbulent planar channel flow. J. Heat Transf. **131**, 061704 (2009)
7. A. Sakurai, K. Matsubara, S. Maruyama, Fundamental study of turbulence-radiation interaction in turbulent channel flow using direct numerical simulation. ASME Paper no. AJTEC2011-44371 (2011)
8. Y.F. Zhang, R. Vicquelin, O. Gicquel, J. Taine, Effects of radiation in turbulent channel flow: analysis of coupled direct numerical simulations. Int. J. Heat Mass Transf. **61**, 654–666 (2013)
9. R. Vicquelin, Y.F. Zhang, O. Gicquel, J. Taine, Effects of radiation in turbulent channel flow: analysis of coupled direct numerical simulations. J. Fluid Mech. **753**, 360–401 (2014)
10. S. Ghosh, R. Friedrich, M. Pfitzner, C. Stemmer, B. Cuenot, M. El Hafi, Effects of radiative heat transfer on the structure of turbulent supersonic channel flow. J. Fluid Mech. **677**, 417–444 (2011)
11. S. Ghosh, R. Friedrich, C. Stemmer, Contrasting turbulence–radiation interaction in supersonic channel and pipe flow. Int. J. Heat Mass Transf. **48**, 24–34 (2014)
12. Y. Wu, D.C. Haworth, M.F. Modest, B. Cuenot, Direct numerical simulation of turbulence/radiation interaction in premixed combustion systems. Proc. Combust. Inst. **30**, 639–646 (2005)
13. Y. Wu, M.F. Modest, D.C. Haworth, A high-order photon Monte Carlo method for radiative transfer in direct numerical simulation of chemically reacting turbulent flows. J. Comput. Phys. **223**(2), 898–922 (2007)
14. K.V. Deshmukh, M.F. Modest, D.C. Haworth, Direct numerical simulation of turbulence–radiation interactions in statistically one-dimensional nonpremixed combustion systems. J. Quant. Spectrosc. Radiat. Transf. **109**(14), 2391–2400 (2008)

15. K.V. Deshmukh, M.F. Modest, D.C. Haworth, Higher-order spherical harmonics to model radiation in direct numerical simulation of turbulent reacting flows. Comput. Therm. Sci. **1**, 207–230 (2009)
16. S. Ghosh, R. Friedrich, Effects of radiative heat transfer on the turbulence structure in inert and reacting mixing layers. Phys. Fluids **27**, 055107–1–14 (2015)
17. S. Ghosh, R. Friedrich, C. Stemmer, LES of turbulence-radiation interaction in plane reacting and inert mixing layers, in *Direct and Large-Eddy Simulation IX*, ed. by J. Frölich, H. Kuerten, B.J. Geurts, V. Armenio (Springer International Publishing, Switzerland, 2015), pp. 489–495.
18. A.A. Townsend, The effects of radiative heat transfer on turbulent flow in a stratified medium. J. Fluid Mech. **3**, 361–375 (1958)

Chapter 6
Turbulence–Radiation Interactions in Atmospheric Pressure Turbulent Flames

6.1 Introduction

As mentioned previously, the importance of radiation in combustion systems has generally, until recently, either been ignored or been treated with simplistic models, such as the optically thin approximation. The realization of the importance of thermal radiation in turbulent flames came concurrently with understanding that radiation was affected by turbulence in much the same way as convection. Consequently, studies on radiation in turbulent flames have all been accompanied by simultaneously considering turbulence–radiation interactions (TRI), and we will also follow this format here. Conversely, thermal radiation can also affect turbulence levels, since its "action at a distance" can smoothen temperature fields. However, the resulting decrease in turbulence is generally very small and has received little attention to date.

TRI exist in both reactive and nonreactive flows. A nonreactive hot mixture of radiatively participating species, typically carbon dioxide and water vapor, may be found in the exhaust sections of almost all combustors, including the regions above fires. Because scalar fluctuations in such nonreactive flows are usually much smaller than in flames, it is commonly believed that the effects of TRI are negligible in such media. Mazumder and Modest substantiated this belief by conducting a series of numerical simulations [1]. An important conclusion of their study was that the role of TRI depends largely on how the temperature fluctuations correlate with the concentration fluctuations of carbon dioxide and water vapor. In most nonreactive flows the fluctuations are found to be uncorrelated and, thus, TRI tend to be negligible. In reactive flows, when a blob of a fuel–air mixture is burnt, it produces high temperatures and high concentrations of carbon dioxide and water vapor locally, resulting in a positive temperature–concentration correlation. Such a positive correlation has a profound impact on radiation calculations, because flame emission is always enhanced (see also Sect. 3.6).

© The Author(s) 2016
M.F. Modest, D.C. Haworth, *Radiative Heat Transfer in Turbulent Combustion Systems*, SpringerBriefs in Applied Sciences and Technology,
DOI 10.1007/978-3-319-27291-7_6

Early attention on flame radiation focussed on radiative intensity emanating from a flame along a single line of sight. In these studies temperatures and concentrations were taken as input data, either calculated from artificial flames or, mostly, from experimental data. Radiative intensities leaving the flame were then calculated in uncoupled fashion using stochastic models to predict turbulence effects. Early results indicated that turbulence always increases the transmissivity of a gas column [2, 3], which was confirmed by experiments. In other early work Fischer et al. [4] performed stochastic simulations to predict intensities leaving an ethanol pool fire, finding that mean radiation intensities were 25–80 % higher when TRI was considered, resulting in better agreement with measurements. The bulk of decoupled line-of-sight radiation calculations has been carried out by the groups of Faeth and Gore, and later Gore et al., e.g., [5–12]. In all cases turbulence resulted in an increase of mean spectral radiation intensities, ranging from small (roughly 10 %) in carbon monoxide/air flames [7], to moderate (about 10–30 %) in methane/air flames [5], to high (of the order of 100 %) in hydrogen/air flames [8]. Similar results were found for luminous flames for continuous radiation from soot. An exhaustive discussion of these studies has been provided by Coelho [13], while we wish to concentrate on coupled CFD/chemistry/radiation studies in this chapter.

6.2 Nonluminous Nonpremixed Jet Flames

A number of experiments have been designed to investigate the structure of nonpremixed flames [14–16]. Among them, a series of piloted-jet methane flames, which were measured by Masri and Bilger at Sydney University [14] and by Barlow and Frank at Sandia National Laboratories [16], have been widely simulated in the literature, and some of these flames have become the target flames of several international workshops on computation of turbulent nonpremixed flames. A bank of measured data of these flames is well documented and is available on the internet [17]. Among these flames, Flame D from Sandia is relatively easy to model, and data for it also include some radiative quantities. For that reason, modelers of radiation in combustion systems have preferentially addressed this flame. More recently, Zheng et al. [18, 19] have carried out extensive radiative measurements on six of the workshop flames. Laboratory flames tend to be small, and small flames only lose small amounts of energy through thermal radiation (radiant fraction $\chi_R \ll 1$) and, therefore, radiation effects are generally ignored in numerical simulations of such flames. Thermal radiation reduces the flames' local temperatures. While the magnitude of that decrease is small in most laboratory flames, it is sufficient to affect the production of some minor species, e.g., nitric oxide (NO).

6.2.1 Sandia D

For the above-mentioned reasons Sandia Flame D has been the most widely studied flame by radiation researchers [20–27]. The basic experimental setup of this flame may be summarized as: the fuel jet ($d_j = 7.2$ mm) with high velocity ($u_j = 49.6$ m/s) is accompanied by an annular pilot flow ($d_p = 18.4$ mm, $u_p = 11.4$ m/s), which is then surrounded by a slow coflow of air ($u_c = 0.9$ m/s). The fuel is a mixture of air and methane with a ratio of 3:1 by volume.

The first ones to study radiation and TRI from basic principles were Li and Modest [20, 21]. They generated a transported PDF turbulence model for the composition variables and attached it to the commercial Fluent CFD code [28] to predict the turbulent flow field with its k–ε model. The simple IEM model was used for turbulent diffusion, and a global reaction rate was used for chemistry. While less accurate than other investigations, temperature and concentration fields were captured reasonably well to assess the effects of radiation on such a small laboratory flame. To study impacts of radiation as well as TRI, the P_1-approximation was employed together with an early version of the FSK spectral model, which considered only CO_2 and H_2O (neglecting CO and CH_4). Three different scenarios were considered. In the first scenario, radiation was completely ignored in order to study the importance of radiation in flame simulations in general. In the second and third scenarios, radiation was considered but TRI were ignored and considered (subject to the OTFA), respectively. The importance of TRI can be assessed by comparing numerical results from the latter two scenarios. By ignoring TRI, it is implied that the two unclosed terms $\langle k \rangle$ and $\langle kaI_b \rangle$ in Eq. (3.96) are evaluated based on the cell means; while by considering it, these two terms are treated exactly. Comparing numerical results the most obvious difference is that the flame gets colder and colder as radiation without TRI and radiation with TRI are considered. Flame peak temperatures of Sandia D were found to drop as a result of considering radiation by 64 K without TRI and an additional 18 K when TRI is included; however, the peak temperature with TRI was, at 2083 K, still more than 100 K above the experimental peak mean value of $\sim$1950 K, mostly due to the simple global reaction model employed. Radiant fractions of 3.1 % (without TRI) and 4.6 % (with TRI) were reported, with an experimentally observed value of 5.1 %.

Two other early studies of Sandia D were reported by Coelho et al. [23] and Coelho [24]. The latter is a post-processing study using experimental temperature and composition data, together with a stochastic turbulence generator. In the first study the flame is simulated using a Reynolds stress model for turbulence modeling, the steady laminar flamelet (SLF) model for combustion (using a 49-species 547-reactions methane mechanism), the DOM for radiation, the SLW model for the radiative properties of the gas, and the OTFA to evaluate the TRI. An assumed beta-PDF was used for turbulent fluctuations of scalar variables. For comparison purposes, calculations were done using 6 different scenarios: (1) no radiation; (2) the optically thin approximation with partial account of emission TRI (employing $\kappa_P(\langle T \rangle, \langle p_\alpha \rangle)\langle I_b(T) \rangle$ for emission); (3) a gray calculation using the DOM and the

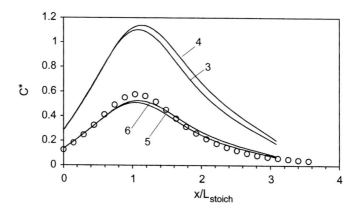

Fig. 6.1 Predicted and measured nondimensional radiant power along the axial direction [23]

same TRI treatment as scenario 2; (4) also gray like 3 but with full TRI (i.e., $\langle \kappa_P I_b \rangle$ for emission and $\langle \kappa_P \rangle$ for absorption); and (5) and (6) being nongray versions of 3 and 4, using the SLW spectral model. Reported temperature drops due to radiation with TRI are similar to those reported by Li and Modest [20, 21], while their calculated radiant fraction with TRI is 5.3 %. In both papers radial heat flux values were also compared against experimental data in the form of the nondimensional radiant power, defined as

$$C^* = \frac{4\pi R^2 q_{\mathrm{rad}}}{S_{\mathrm{rad,exp}}}, \tag{6.1}$$

where q_{rad} is the radial radiative heat flux, which is a function of axial position, and $R = L_{\mathrm{stoich}}/2$ is the radius where the heat flux gauge is located, with L_{stoich} being the stoichiometric flame length. C^* is nondimensionalized by the measured total radiant power, $S_{\mathrm{rad,exp}}$. C^* cannot be determined with the optically thin approximation and, as seen from Fig. 6.1, the use of a gray model (with or without TRI) greatly overpredicts the heat loss from the flame (almost equal to total emission, i.e., nearly equal to the optically thin approximation). Considering nongray radiation, however, predicts substantial self-absorption, with full TRI (scenario 6) resulting in slightly less absorption than the one neglecting the absorption coefficient–Planck function correlation (5). Both nongray predictions show excellent agreement with the experimental results.

The first ones to model radiation and full TRI (i.e., not subject to the OTFA) were Wang et al. [25], employing the LBL-accurate PMC for stochastic particles, which was specifically developed for this purpose (see description in Sect. 3.5.3). Similar to Li and Modest [20, 21] they used a transported composition PDF attached to a CFD code. However, a different CFD code, STAR-CD [29], a more advanced composition PDF code and a more realistic chemical reaction mechanism involving 16 species and 41 steps [30] versus their single-step mechanism were employed.

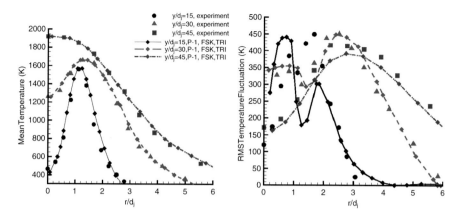

Fig. 6.2 Radial profiles of mean temperature (*left*) and RMS temperature fluctuations (*right*) at various axial locations of Flame D [27]

In addition, in order to match the experimental data better, the value of $C_{\varepsilon 1}$, a constant in the k–ε model, was chosen to be 1.48 rather than the standard value of 1.44 (the same value was also employed for scaled flames discussed in the next section). They observed similar temperature drops for no-radiation, to nongray without TRI, and to nongray with partial and full TRI. Their prediction of a maximum temperature (as well as of the entire temperature field, cf. Fig. 6.2) was essentially equal to the experimental values. While expected, they proved that the OTFA is indeed valid for such small flames, i.e., the effects of absorption TRI were found to be negligible.

Finally, in a study aimed at comparing the usefulness of various RTE solvers and spectral models, Pal et al. [27] also modeled Flame D using the same flow solver, showing that for such a flame with little radiation relatively simple radiation treatment suffices—as long as nongray properties (and, to a lesser extent, TRI) are accounted for. Figures 6.2 and 6.3 compare mean and rms values of temperature and NO mass fraction, as calculated by Pal et al. using a P_1 solver combined with the FSK spectral model, against experimental data. It is seen that mean temperature values are captured extremely well by their model, and that even the rms fluctuations of temperature are predicted reasonably well. NO mass fractions are shown in Fig. 6.3 because NO values are extremely sensitive to even small temperature changes, making them an excellent tool to study the importance and accuracy of radiation models. It is observed that mean values of NO, and even their rms fluctuations, are predicted very well if radiation and TRI are considered, while—even in this small flame with little radiation—neglecting radiation results in vast overprediction of NO and its rms values.

In [26], an OpenFOAM-based fully coupled LES/PDF/LBL–PMC solver was developed to simulate turbulent flames with radiative heat transfer. Development and validation were carried out using Sandia Flame D as the modeling target. An unstructured mesh of approximately 1.2 million hexahedral cells was used in

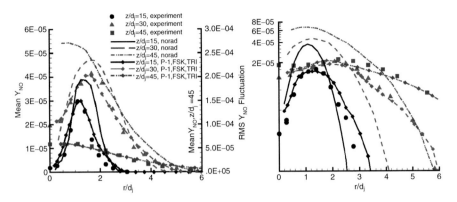

Fig. 6.3 Radial profiles of mean NO mass fraction (*left*) and RMS mass fraction fluctuations (*right*) at various axial locations of Flame D [27]

a cylindrical computational domain that extended 70 fuel-jet diameters in the axial direction and 18 fuel-jet diameters in the radial direction, with approximately 15 particles per cell for the PDF method. A digital-filter-based turbulence synthesis technique was used to generate realistic turbulence at the inlet, and a 16-species/ 41-reaction skeletal methane/air chemical mechanism was used for chemical kinetics. Approximately two million photon bundles were traced on every computational time step, and computational parallelization strategies were proposed and tested. For comparison purposes, results from a one-step global chemical mechanism, results from a RANS/PDF method, and results for finite-volume chemistry and radiation (neglecting influences of subfilter-scale fluctuations on the resolved fields) were also presented. Results with the more complete chemical mechanism and with consideration of subfilter-scale fluctuations were in better agreement with experimental measurements (Fig. 6.4). The RANS/PDF results in Fig. 6.4 are not in as good agreement with experiment as those shown earlier, because of the less accurate (and less computationally intensive) chemical mechanism that was used here.

6.2.2 Artificial/Scaled flames

Recognizing that small laboratory flames, such as Flame D, have little radiation and, thus, radiation has only limited impact on the flame's characteristics, a number of researchers around the groups of Modest and Haworth artificially scaled Flame D up in size to make radiation effects more pronounced. In their study, Li and Modest [21] investigated the effects of traditional nondimensional parameters, such as Reynolds number (based on fuel-jet diameter and velocity), Damköhler number (the ratio of flow-to-reaction time scales), Froude number (buoyancy effects), and optical thickness of the flame, defined as

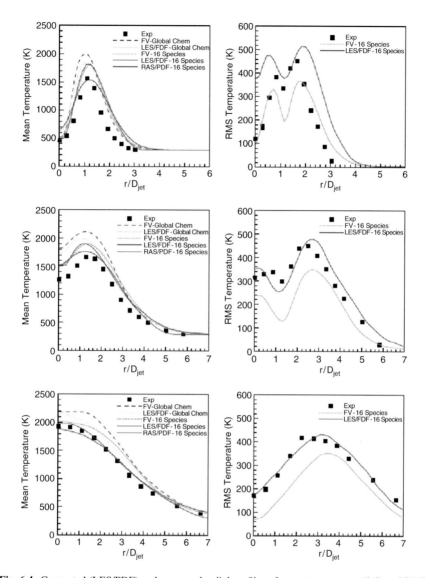

Fig. 6.4 Computed (LES/PDF) and measured radial profiles of mean temperature (*left*) and RMS temperature fluctuations (*right*) at three axial locations (*top to bottom*: $x/D_{jet} = 15$, $x/D_{jet} = 30$ and $x/D_{jet} = 45$) for Sandia Flame D [26]. "FV" refers to simulations that ignore subfilter-scale fluctuations

$$\tau = \kappa_P L, \tag{6.2}$$

where κ_P is an average Planck-mean absorption coefficient of the participating medium, i.e., the combustion products of H_2O and CO_2, and L is the flame length. For turbulent jet flames, flame length is approximately a linear function of jet

Table 6.1 Summary of radiation calculation results for series of κL-flames [21]

	No rad.	Without TRI				With TRI				$\frac{f^2 - f^1}{f^1}$
	T_{max}	T_{max}	$\dot{Q}_{em}$	$\dot{Q}_{net}$	f^1_{rad}	T_{max}	$\dot{Q}_{em}$	$\dot{Q}_{net}$	f^2_{rad}	
Flame	(K)	(K)	(kW)	(kW)	(%)	(K)	(kW)	(kW)	(%)	(%)
$\kappa L.1$	2165	2101	0.624	0.534	3.05	2084	0.928	0.798	4.56	49
$\kappa L.2$	2161	2016	4.12	2.98	8.51	1952	5.33	3.92	11.2	32
$\kappa L.3$	2169	1842	21.68	12.12	17.3	1725	20.94	12.68	18.1	4.6

diameter [31] and, in their study, was estimated to be $L = 40d_j$. The base flame was Sandia's Flame D, for which they determined an optical thickness of 0.237. They considered two other (artificial) flames derived from Flame D by doubling and quadrupling the jet diameter, with corresponding flame optical thicknesses of 0.474 and 0.948, respectively, naming them $\kappa L.1$, $\kappa L.2$, and $\kappa L.3$. In order to study isolated effects of optical thickness, they kept Re constant (by cutting velocities by factors of 2 and 4) as well as Da (by cutting reaction rates by factors of 4 and 16). Three different scenarios were considered for each flame: in the first radiation is completely ignored and, in the second and third scenarios, radiation is considered but TRI are ignored and considered (but subject to the OTFA), respectively. As already indicated for Flame D, the most obvious difference between the three scenarios is that the flame gets colder and colder as radiation without TRI and radiation with TRI are considered. This is universally true for every flame although the trend is more obvious for flames with large optical thickness. Flame peak temperatures for different flames are tabulated in Table 6.1. While the peak temperature drops only 64 K and an additional 18 K for the small optical thickness Flame D, it drops by 145 and 64 K, respectively, for a medium flame, and by 327 and 117 K for a large optical thickness flame. Although peak temperature applies only to a single point, it usually characterizes the entire temperature field. Clearly, ignoring radiation leads to severely overpredicted flame temperatures, especially for large flames, such as Flame $\kappa L.3$. Moreover, TRI account for about one third of the total temperature drop due to radiation.

The most important quantity that describes the overall radiation field of a flame is the net radiative heat loss ($\dot{Q}_{net}$) from the flame, and its normalized variable, the "radiant fraction" (χ_R). These quantities are also included in Table 6.1. As the flame's optical thickness is increased, the flame radiant fraction increases quickly and the flame gets colder as discussed earlier. In the study of Li and Modest [21] optical thickness was varied by changing the size of the flame, keeping Re fixed (i.e., lowering velocities). Thus, as the flame gets larger, the flow residence time becomes longer, which implies that an average fluid particle will lose more energy through radiation. As a result, the radiant fraction increases as flame size increases. The radiant fraction is only about 5 % for Flame $\kappa L.1$, but as high as 18 % for Flame $\kappa L.3$. This also explains why temperature levels drop more significantly in optically thick flames. The table also shows how the TRI enhance radiative heat transfer. For Flame $\kappa L.1$, the net radiative heat loss from that flame is increased from 0.534 to 0.798 kW, indicating a 49 % increase as a result of TRI. In contrast, total radiative heat loss increases by 32 % for Flame $\kappa L.2$, but only 4.6 % for Flame $\kappa L.3$ is due to TRI. As the flame gets optically thicker, the actual values of radiative heat loss, ignoring TRI and considering TRI, become closer and closer. This does not mean that considering TRI is less important for optically thick flames. Radiation calculations are strongly dependent on the flame temperature level, which has greatly decreased as a result of TRI. Thus, comparison of the radiative loss quantities alone would be misleading.

The role of TRI on radiative heat transfer can be better understood by isolating their effects on the radiation calculations alone. This can be done by freezing a snapshot of the turbulent flow field (stochastic particles' species concentrations and their temperatures in a transported PDF study) at a point in time, and then calculating radiation fields by ignoring and considering TRI, respectively. Since the same flow field is used, any differences in the results are caused entirely by TRI. The frozen study can also help to differentiate, which correlations making up the full TRI are the most important. Seven different scenarios were considered and are summarized in Table 6.2, where quantities evaluated from mean composition variables are denoted with an overbar. Cases TRI-1 and TRI-2 show the nonlinear effects of the absorption coefficient on absorption and combined emission–absorption. Nonlinear temperature dependence of the absorption coefficient of combustion products tends to increase, both, absorption and emission, but only by a small amount, i.e., 10 % or less. TRI-3 shows the effects of the temperature self-correlation, having the strongest impact on emission (an increase of roughly 35 % for all flames). This is misleading, however, since combustion gases are highly nongray: TRI-4 shows that the influence of the nongray Planck function self-correlation, $\langle aI_b \rangle$ (with the inclusion of the nongray stretching factor a leading to selective integration of $I_{b\eta}$ over the spectrum) is rather small, generally less than 10 %. For nongray gases the main contribution to TRI comes from the absorption coefficient–Planck function correlation, as seen from TRI-5 and TRI-F, leading to increased emission

Table 6.2 Comparison of TRI components for series of κL-flames [21]

Flame		TRI-N	TRI-1	TRI-2	TRI-3	TRI-4	TRI-5	TRI-F
Emission		$\bar{k}\bar{a}\bar{I}_b$	$\bar{k}\bar{a}\bar{I}_b$	$\langle k \rangle \bar{a}\bar{I}_b$	$\bar{k}\bar{a}\langle I_b \rangle$	$\bar{k}\langle aI_b \rangle$	$\langle k\bar{a}\bar{I}_b \rangle$	$\langle k\bar{a}\bar{I}_b \rangle$
Absorption		$\bar{k}$	$\langle k \rangle$	$\langle k \rangle$	$\bar{k}$	$\bar{k}$	$\bar{k}$	$\langle k \rangle$
$\kappa L.1$	$\dot{Q}_{em}$ (kW)	0.597	0.597	0.657	0.812	0.641	0.928	0.928
	$\dot{Q}_{net}$(kW)	0.516	0.507	0.558	0.706	0.555	0.820	0.798
	f (%)	2.94	2.90	3.19	4.03	3.17	4.68	4.56
$\kappa L.2$	$\dot{Q}_{em}$ (kW)	3.42	3.42	3.73	4.59	3.69	5.33	5.33
	$\dot{Q}_{net}$(kW)	2.51	2.45	2.68	3.42	2.73	4.02	3.92
	f (%)	7.14	6.98	7.64	9.74	7.78	11.5	11.2
$\kappa L.3$	$\dot{Q}_{em}$ (kW)	12.7	12.7	13.8	17.5	14.2	20.9	20.9
	$\dot{Q}_{net}$(kW)	7.63	7.44	8.03	10.6	8.63	13.1	12.7
	f (%)	10.9	10.6	11.5	15.1	12.3	18.7	18.1

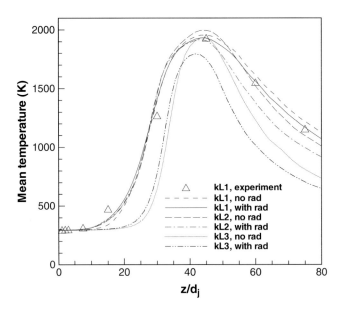

Fig. 6.5 Centerline profiles of mean temperature in considered flames [25]

of approximately 60 % for all flames. Therefore, one can conclude that, while for the (very unrealistic case) case of a gray gas, it is the temperature self-correlation that dominates TRI, in a realistic nongray gas, it is the absorption coefficient–Planck function correlation that contributes most to TRI.

Wang et al. [25] and Pal et al. [27] also modeled two and fourfold scaled-up versions of Flame D, again keeping Re constant. However, they used the same realistic 16 species–41 step methane mechanism for all 3 flames (rather than keeping Da constant, it increases by a factor of 4 and 16, respectively). The centerline mean temperature profiles evaluated with and without radiation are presented in Fig. 6.5 for all three flames. As discussed earlier it is seen that radiation has a cooling effect, making the flames shorter. The impact of radiation, implied by the difference between profiles with and without radiation, becomes more and more significant as the flame optical thickness increases from $\kappa L.1$ to $\kappa L.3$.

Wang et al. [25] also investigated the effects of TRI on the 3 flames. Their stochastic particle LBL PMC solver remains the only one to date able to model TRI without applying the OTFA. Table 6.3 shows total flame emission and radiative heat loss for the cases on no TRI, TRI subject to the OTFA, and full TRI. It is observed, as is expected for nonluminous flames, that radiative losses with and without the OTFA are virtually identical, i.e., it is safe to neglect absorption TRI for such flames. It is also seen that the increase of radiative loss for Flame D is somewhat smaller (30 %) than the one predicted by the less accurate analysis of Li and Modest ($\kappa L.2$ and $\kappa L.3$ are different and should not be compared).

Table 6.3 Total emission and net radiative heat loss using different TRI treatments (in kW) for 3 scaled flames [25]

Flames	No TRI		TRI with OTFA			Full-TRI		
	$\dot{Q}_{em}$	$\dot{Q}_{net}$	$\dot{Q}_{em}$	$\dot{Q}_{net}$	$\varDelta$TRI (%)	$\dot{Q}_{em}$	$\dot{Q}_{net}$	$\varDelta$TRI (%)
$\kappa L.1$	0.849	0.500	1.07	0.648	29.6	1.07	0.645	29.0
$\kappa L.2$	5.01	2.57	6.22	3.29	28.0	6.24	3.27	27.2
$\kappa L.3$	17.44	8.48	22.94	11.57	36.4	22.95	11.37	34.1

Pal et al. [27] used essentially the same chemistry and composition PDF solver, but embedded into the open-source CFD solver OpenFOAM [32]. They focussed on evaluating the accuracy and efficiency of different radiation models (RTE solvers as well as spectral models). They also considered the effects of soot radiation and, therefore, their results are discussed in the following section on luminous flames. A LES/PDF modeling study of the same flames (focusing on the relative importance of resolved-scale fluctuations versus subfilter-scale fluctuations to TRI) [33] is also discussed there.

6.3 Luminous Nonpremixed Jet Flames

Few studies appear to have been conducted to date to evaluate the importance of radiation and of TRI in luminous flames. Since soot emits and absorbs radiation over the entire spectrum, it may be expected that radiation and TRI would be particularly important in luminous flames. And, since heavily sooting flames can be optically rather thick, the validity of the OTFA (i.e., neglect of absorption TRI) may be questionable. Perhaps the first to research feedback of radiation and TRI onto the flame were Adams and Smith [34], who investigated a 3D natural gas fired industrial furnace, using a fairly basic mix of models (RANS CFD with a k–ε turbulence model for the flow field, a statistical mixture fraction method to predict the joint PDF for local instantaneous stoichiometry and enthalpy for turbulent combustion, and an empirical soot formation–destruction model). Both soot and gas were modeled as gray, the RTE was solved with the discrete ordinates method, and the OTFA was applied. They compared 4 different scenarios, i.e., cases without and with soot, as well as without and with TRI. They noted correctly that, in the absence of soot, TRI increases emission to some degree despite the concurrent drop in temperature levels. Considering soot increased emission 10- to 20-fold if TRI are neglected, but by a much smaller amount in the presence of soot, probably caused by strongly decreasing soot levels as predicted by their simple algebraic model.

Tessé et al. [35] modeled an open diffusion flame, which was experimentally studied by Coppalle and Joyeux [36], in which pure gaseous ethylene is injected vertically upward in atmospheric air. They used a RANS CFD code with a k–ε turbulence model for the flow field, as well as simplified chemistry and radiation models, in order to match temperature and velocity profiles of the experiment. Then they applied a particle Monte Carlo method to obtain the 3D PDF of the reaction progress variable, mixture fraction and soot volume fraction, by tracking fluid packets through the mean flow field. Gas chemistry was described by 38 species and 119 reactions, while five parameters were used to describe the soot particle behavior. They assumed that the thermophysical properties of individual coherent turbulent structures were uniform and randomly obtained from the 3D PDF to construct a joint composition PDF of temperature and concentrations for each individual turbulent structure. Finally, a Monte Carlo ray tracing scheme was carried out, in which ray paths were divided by turbulent structures into continuous segments, in which the radiative properties were evaluated from the corresponding joint PDF using the narrow band k-distribution database for CO_2 and H_2O as collected by Soufiani and Taine [37]. Their scheme was the first to take absorption TRI into account without invoking the OTFA. However, the composition PDF was obtained decoupled from the flow and temperature fields, rather than by solving the composition PDF transport equation using the particle Monte Carlo method directly. They showed that for this flame both emission and absorption (and, thus, the radiative source) increase by roughly 30 % due to TRI (i.e., much less than the prediction of Adams and Smith [34], and with opposing trends). This increase comes from substantially increased emission away from the centerline, while emission at the centerline itself decreases. However, it appears that in this study no feedback to the flame is considered, i.e., radiative sources with and without TRI are based on the same temperature and concentration field. Also, no indication of the relative importance of absorption TRI was given.

In addition to the scaled Sandia D flames discussed in the previous section, Pal et al. [27] also considered a luminous flame by adding soot according to a simple state relationship, with a maximum soot volume fraction of about 6 ppm. Table 6.4 shows some of their results for Flame D scaled 4 times and with soot added. It is seen that, in this optically thicker flame, radiation causes temperatures to drop by almost 160 K, while TRI increases heat loss by 65 % and brings upon an additional drop of 150 K. This has tremendous impact on NO production, with the NO index falling by factors of 25 and an additional 6, respectively. Looking at RTE solvers, one sees that P_1 overpredicts net heat loss by about 20 %, while P_3 and the FVM/DOM with 16×4 ordinates are within 2 % or so. Also included in the table are results of the more advanced multi-scale FSK method (MSFSK) [38], which returns near-LBL results. Differences between the two methods indicate that the basic FSK underpredicts heat loss by roughly 5 %.

The CPU times reported refer to simulations performed on Triton, a PC Cluster run by the High Performance Computing Group of the University of California, San Diego. The cluster consists of 256 nodes, each of which has four 3.0 GHz Intel Nehalem E5530 dual-core processors. Times reflect calculations on a 52,800 cell wedge mesh for one time step (averaged over 1000 time steps) on a single processor. 16-processor parallel computations demonstrated an approximately 80 % parallel efficiency for, both, radiation and chemistry calculations, with the latter being responsible for the bulk of CPU use. For the P_N solutions values from previous

Table 6.4 Effects of radiation model and TRI on simulation of luminous Flame 4×D+Soot
$[t_{rad} = t_{prop} + t_{RTE}, t_{total} = t_{flow} + t_{chem} + t_{prop} + t_{RTE}]$ [27]

RTE Solver	Spectral model	TRI	Peak mean T (K)	NO emission index (g_{NO}/kg_{fuel})	$\dot{Q}_{net}$ (kW)	% ΔTRI	t_{prop} (s)	t_{RTE} (s)	t_{rad} (s)	t_{total} (s)
Off			2011	7.50						72.70
PMC	LBL	No	1854	0.30	18.5					
		Yes	1701	0.05	30.6	65			3.20	75.92
Optically Thin		No	1701	0.06	30.8					
		Yes	1588	0.001	50.7	65				
P-1	FSK	No	1821	0.21	21.6					
		Yes	1690	0.03	35.1	63	0.32	0.10	0.42	73.14
	MSFSK	No	1813	0.19	22.9					
		Yes	1688	0.08	37.2	62	0.64	0.16	0.80	73.53
P-3	FSK	No	1860	0.36	17.6					
		Yes	1704	0.05	29.1	65	0.32	1.05	1.37	74.09
	MSFSK	No	1858	0.35	18.0					
		Yes	1702	0.05	29.6	64	0.64	2.00	2.64	75.36
FVM(16×4)	FSK	No	1848	0.32	19.6					
		Yes	1693	0.04	32.1	63	0.32	6.71	7.03	79.71
	MSFSK	No	1850	0.34	18.7					
		Yes	1699	0.05	31.0	66	0.64	13.39	14.03	86.75

time steps were used to speed up iteration; for PMC 80,000 photon bundles were traced during a single time step, which were time-blended over 100 time steps. It is seen that assembly of k-distributions dominates radiation time requirements for P_1 and P_3; The FVM requires 16×4 ordinates to achieve P_3 accuracy, but uses substantially more CPU time than P_3.

The most sophisticated model of luminous flames to date was presented by Mehta et al. [39, 40], who employed the hybrid RANS/transported PDF scheme used by Wang et al. [25] for flow and composition variables, combined with Wang's LBL-accurate PMC to evaluate radiation and TRI without the need of applying the OTFA (cf. Sect. 6.2.1). The modeling suite was complemented with a state-of-the-art method-of-moments soot model [41], using a systematically reduced 33-species reaction mechanism containing 205 elementary reactions to model gas-phase kinetics [42]. This mechanism was found to perform satisfactorily in predicting soot volume fractions in laminar premixed and opposed-flow-diffusion ethylene–air flames across a broad range of conditions. Nongray soot properties based on the correlation of Chang and Charalampopoulos [43] were added to the PMC module. In [39] the emphasis is on comparison of the full model against experimental results from six flames: 2 ethylene flames [36, 44] and four oxygen-enriched methane–ethylene flames [45]. An example is given in Fig. 6.6, showing measured and predicted centerline temperatures and soot volume fractions for the Coppalle and Joyeux flame [36], demonstrating very good agreement. Not surprisingly, neglecting the strong soot emission would lead to overprediction by hundreds of degrees, while the optically thin model leads to underpredicted temperatures. These underpredicted temperatures then also lead to underpredicted soot levels, which are captured well if accurate radiation is included.

In [40] Mehta et al. investigated the influence of TRI and its components (in particular, absorption TRI) on luminous flames. Fully coupled solutions with and

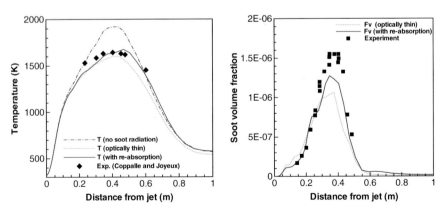

Fig. 6.6 Measured (*symbols*) and computed (*lines*) centerline mean temperature (*left*) and soot volume fraction (*right*) profiles for Coppalle and Joyeux flame. Computed profiles are shown for different levels of radiation modeling [39]

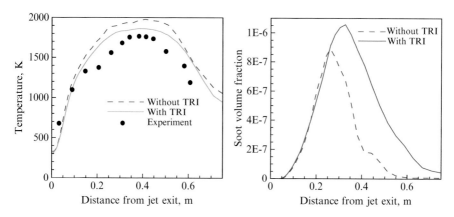

Fig. 6.7 Comparison between model predictions for the centerline mean temperature and soot volume fractions for Kent and Honnery flame, with and without TRI [40]

without TRI result in different flame structures, which feed back into the flow solver through the mean density field. To investigate the overall influence of TRI on the sooty flames, simulations were also carried out with radiative properties evaluated based on cell mean values of concentrations and temperature. Computed and measured centerline mean temperature profiles for the Kent and Honnery flame [44] ("Flame II") are shown in Fig. 6.7. Neglecting TRI increases the peak temperature by 75 K, and the computed peak centerline mean soot volume fraction is reduced from 1.2 ppm to just over 0.8 ppm. The flame is hotter when TRI is not included while the net radiative loss is lower nonetheless. For the 4 oxygen-enriched flames (Flames III–VI) radiative wall flux data were available, which compared well with the predicted ones (with TRI), while no-TRI fluxes were 10–50 % lower. This observation (valid for all flames studied) is in direct contrast to the work by Adams and Smith [34] (perhaps due to their rudimentary soot and radiation models). On the other hand, higher temperatures without TRI do not necessarily translate to lower soot levels: in all of the Endrud flames [45] the opposite trend was observed.

To systematically isolate TRI effects, Mehta et al. [40] also carried out a frozen field study. The PMC method was used in conjunction with the frozen fields consisting of stochastic particles carrying all scalar variables (as generated by the transported PDF turbulence method). Full TRI (emission and absorption both based on particle values), partial TRI (only emission is based on particle values), and no TRI (emission and absorption are both based on cell mean quantities) are considered, to estimate the corresponding radiative transfer from the flames. The results of different TRI treatments on volume-integrated quantities characterizing Flames I–VI are summarized in Table 6.5. Net emission increases by $\sim$30–60 % when accounting for TRI [i.e., the increase of $\langle \kappa_P I_b \rangle$ over $\kappa_P(\langle T \rangle, \langle p_\alpha \rangle)I_b(\langle T \rangle)$]. Absorption TRI values are shown as a percentage of absorption with TRI (i.e., the change of $\langle \kappa_\eta G_\eta \rangle$ vs. $\kappa_P(\langle T \rangle, \langle p_\alpha \rangle)\langle G_\eta \rangle$). Absorption TRI is always negligible for the laboratory-scale flames I–VI. However, it is expected that absorption TRI may be

Table 6.5 Emission and absorption TRI characteristics for various sooting flames (frozen-field analysis)

Flame	Fuel	Oxidizer	Ref.	Emission TRI (%)	Absorption TRI (%)	Emission (%) Soot ÷ Gas	Reabsorption (%) Soot	Reabsorption (%) Gas
I	Ethylene	Air	[36]	39	0.4	39 ÷ 61	8	54
II	Ethylene	Air	[44]	57	0.1	43 ÷ 57	7	45
III	Blend	Air	[45]	41	0.3	5 ÷ 95	3	46
IV	Blend	30 % O_2	[45]	32	1.0	9 ÷ 91	3	42
V	Blend	40 % O_2	[45]	44	0.2	7 ÷ 93	3	41
VI	Blend	55 % O_2	[45]	38	0.6	4 ÷ 96	1	35
VII	Ethylene	Air		20	−6	68 ÷ 32	74	86

Blend: 90 % methane–20 % ethylene; VII: Flame II scaled 32×

significant in sufficiently optically thick flames. Since Flame II yielded the highest soot volume fractions of the flames considered, Mehta et al. [40] also simulated a scaled-up Flame II with jet diameter a factor of 32 larger than Flame II. Results from a frozen field analysis for the large flame (Flame VII) are also included in Table 6.5. Soot emission is more than twice that of the gas-phase emission, even though the sooting region is much smaller than that of the participating gases. Almost 85 % of all the gas-phase emission is reabsorbed in the computational domain. This is significantly higher than the gas-phase reabsorption in laboratory-scale flames, because the optical thickness of the entire system is much higher than previously encountered. Similarly, almost 75 % of the soot emission is reabsorbed, compared to a maximum of approximately 10 % in laboratory-scale flames. It is also observed that emission TRI decreases, while absorption TRI at −6 % (−10 % for the soot) is no longer negligible. Negative absorption TRI is expected, since it is known that transmissivities increase due to TRI. The results indicate that absorption TRI may be neglected in relatively small flames, irrespective of whether they are sooting flames or not. For large sooting flames, on the other hand, absorption TRI may be small but noticeable. A similar study to [40] was carried out a little earlier by Wang [46], who studied the ethylene flame of Lee [47] as well as a scaled-up version, using a simplistic state relationship for soot. He came to the same overall conclusions, but found nonnegligible local absorption TRI effects even for the small flame; however, this may have been due to the primitive soot model used, since it was only observed when feedback was considered (i.e., *not* in the frozen field study).

LES/PDF results for the same four Sandia-D-based flames that were studied in [27] (Sandia D, Sandia D with artificial soot, Sandia D×4, Sandia D×4 with artificial soot) were presented in [33]. Computed instantaneous snapshots of the resolved temperature and velocity fields for Flame D are shown in Fig. 6.8. The focus of the study was to explore TRI in LES of luminous and nonluminous nonpremixed jet flames, and in particular, to quantify the relative contributions of resolved-scale fluctuations versus subfilter-scale fluctuations to emission and absorption TRI. The simulations feature a transported PDF method for subfilter-scale fluctuations in composition and temperature, and a fully coupled PMC method for radiative

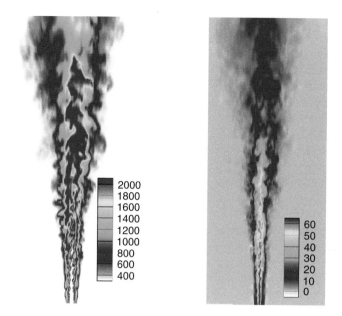

Fig. 6.8 Instantaneous snapshots of resolved temperature (*left*) and axial velocity (*right*) fields on a cutting plane through the geometric axis of symmetry for Flame D. Figures courtesy of Dr. Ankur Gupta

transfer with line-by-line spectral resolution. The model was exercised to isolate and quantify individual contributions to TRI for conditions that ranged from small optically thin flames to relatively large optically thick flames, including spectral molecular gas radiation and broadband soot radiation. The results provided new physical insight into TRI and guidance for modeling. In all cases, emission TRI were found to be responsible for a significant fraction of the radiative emission, and that fraction increased with increasing optical thickness. For simulations where approximately 84 % of the turbulence kinetic energy was resolved, contributions of subfilter-scale fluctuations to emission TRI exceeded those of resolved-scale fluctuations. The largest contributions to emission TRI were the absorption coefficient–temperature correlation and the temperature self-correlation. Absorption TRI were evident only for relatively high optical thicknesses. In all cases, the contributions of subfilter-scale fluctuations to absorption TRI were negligible.

6.4 Pool Fires

Pool fires are buoyancy driven turbulent diffusion flames, and are most often representative of unwanted fires. By necessity, even at laboratory scale they tend to be a bit larger than the small jet flames. As with other combustion systems, the impact of thermal radiation on pool fires has been commonly ignored or treated with simple optically thin or gray models. Fischer et al. [4] measured exiting radiation from a 0.5 m diameter ethanol pool fire, finding strong soot radiation and a radiant fraction of 19 %. They also carried out some calculations based on the measured mean temperature, rms values of temperature fluctuations, and species concentration coupled with a prescribed pdf of the temperature. They found that mean radiation intensities along the axis were 25–80 % stronger when TRIs were considered. Klassen and Gore [48, 49] drew similar conclusions from measurements and calculations for toluene-fired pool fires, stating that ignoring TRI for soot may underpredict intensities by an order of magnitude. The first computational study of a pool fire was the one by Snegirev [50], simulating propane pool fires experimentally studied by Gengembre et al. [51, 52]. They employed their Fire3D code with a k–ε model (tuned for better agreement with experiment), and a simple chemistry scheme. Radiation was determined with a Monte Carlo method using gray properties, and also the weighted-sum-of-gray-gases [53]. TRI was also considered in a simplified manner (for the gray case): fluctuations of species concentration were ignored and the absorption coefficient and the emission term were expanded into a Taylor series and truncated with two constants adjusted to match experimental radiant fractions and radiative fluxes on remote targets.

State-of-the-art simulations of small pool fires have been carried out by Consalvi [54] and Consalvi et al. [55, 56]. The fire-induced flow was modeled using a buoyancy-modified k–ε model and the SLF model coupled with a presumed probability density function (pdf) approach, with transport equations for the turbulent kinetic energy, the dissipation rate of turbulent energy, the mixture fraction, the mixture fraction variance, the soot mass fraction, the soot number density per unit mass of mixture, and the total enthalpy. Thermal radiation was determined with a DOM/FVM RTE solver paired with a high-level spectral model. In [56] different spectral models were tested, and the FSK method was established as the best, which was then used in the remaining work. In [54] a 34-kW methane pool fire produced by a burner with a diameter of 0.38 m was simulated, which had been tested experimentally by Hostikka et al. [57], while in [55, 56] also the propane flames of Gengembre et al. [51, 52] were modeled. Figure 6.9 compares their calculated outward radiative fluxes for two propane pool flames with experimental data (at a distance d from the burner axis), and also with the results of Snegirev [50]. Agreement with experiment is seen to be quite good (and much better than the results of Snegirev). For the methane pool flame, like other researchers they also determined the influence of radiation and of TRI on temperature fields. In this strongly sooting fire temperature drops due to radiation were assessed to be as high as 300 K near the maximum mean temperature, with up to an additional 150 K due to TRI in some locations. They also calculated the contributions of different TRI components, using the full-TRI flowfield (subject to the OTFA) as a frozen field, as shown in Fig. 6.10. They broke up partial TRI differently from Li and Modest [21] for nonluminous jet flames. However, their results are consistent: all parts of TRI increase radiative heat loss, and only neglecting the absorption coefficient self-correlation (Mod5, equal to TRI-5 in Table 6.2) leads to higher heat loss than full-TRI (Mod1 = TRI-F), because it underpredicts self-absorption.

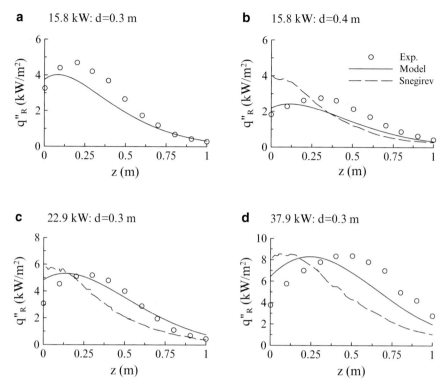

Fig. 6.9 Vertical distribution of outward radiative flux; model predictions (*solid lines*) are compared with experimental data (*open circles*) [51, 52]; numerical predictions obtained by Snegirev [50] are also plotted (*dashed lines*); from [55]

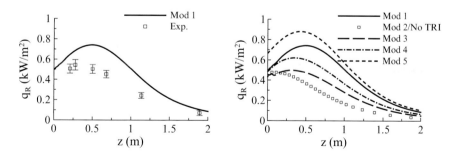

Fig. 6.10 Vertical distribution of outward radiative flux at a distance of 0.732 m from axis of flame [54]. *Left*: model predictions (*solid lines*) are compared with experimental data (*open symbols*); *right*: influence of different TRI closures (*Mod1* TRI-F, *Mod2* TRI-N, *Mod3* TRI-4, *Mod5* TRI-5)

6.5   Pulverized Coal and Oxy-Fuel Combustion

Electricity generation relies heavily on coal-fired plants, a carbon-intensive fossil fuel. Strategies to reduce CO_2 emissions from coal-fired power plants include pre-combustion capture, post-combustion capture, and oxy-fuel combustion. In oxy-fuel combustion, a mixture of oxygen and recycled flue-gas replaces air as the oxidizer to produce products that contain a high concentrations of CO_2, which facilitates the separation of CO_2 from the other constituents. Flue-gas recirculation is usually used to maintain peak temperatures and heat transfer rates that are comparable to those in a conventional air–fuel combustion system, thus minimizing combustor design changes.

The simulation of a coal–air flame requires a multiphase flow model with separate energy equations for dispersed phase and the carrier air flow. The air flow is commonly modeled as a continuum medium in a Eulerian framework, while the dispersed coal particles may be modeled either in a Lagrangian or Eulerian framework. The former approach tracks discrete particles on a mesh in a Lagrangian fashion, while the latter treats particles as an inter-penetrating continuum governed by Eulerian transport equations. Both approaches have their intrinsic advantages. Since radiation in standard air-fired coal combustion systems is dominated by near-gray coal particles, most simulations to date have used simple T^4-relationships. Exceptions are the recent papers of Cai et al. [58, 59]. In the first paper the Eulerian–Eulerian open-source Multiphase Flow with Interphase eXchanges (MFIX) [60] solver was employed to model a small laboratory burner [61]. In the other paper OpenFOAM [32] was used with its Lagrangian dispersed phase model, again to model a small laboratory burner [62]. In both cases the FSK method was used for radiation, augmented by a regression scheme to accommodate nongray coal particles [63], together with a P_1 RTE solver. While both systems are relatively small and optically thin, and radiation is dominated by emission from gray coal particles, nongray effects were found to be nonnegligible, raising particle temperatures by about 100 K (and, to a lesser extent, gas temperatures), as shown in Fig. 6.11. As is generally the case for gas radiation, gray and optically thin analysis return essentially identical results here for the coal case as well, indicating that all coal emission escapes from the flame, while a substantial fraction of the nongray gas emission is self-absorbed. Disparity between experiment and simulation results also points out the great uncertainties that still accompany multiphase reactive flow modeling.

During oxy-fuel combustion the increase in the concentration of radiatively participating species, mostly CO_2 and H_2O, significantly increases radiative heat transfer rates. Coal particle radiation no longer dominates over gas radiation, and nongray effects are expected to become very important. While the importance of radiation has been recognized in early oxy-fuel analyses, its treatment was generally limited to gray models included in CFD packages, such as ANSYS Fluent and others [64]. One exception was Nikolopoulos et al. [65], who added the (today outdated) exponential-wide-band-model [66] to Fluent and its built-in DOM RTE

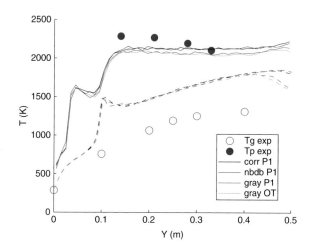

Fig. 6.11 Axial temperature for laboratory coal flame; comparison between experiments [62] and simulations [59]. *Dots* are measurements, *solid lines* are predictions of particle pyrometer readout, *dashed lines* are gas-phase centerline temperature (*OT* optically thin radiation, *nbdb* FSK from narrow band database, *corr* FSK from algebraic correlations)

solver without, however, discussing the impact of radiation. The authors are aware of three recent numerical studies that use state-of-the-art radiation models in oxy-fuel applications.

Zhao et al. [67] modeled the OXYFLAM-2A oxy-natural gas flame, a coaxial nonpremixed jet flame issuing into a refractory-lined furnace, which is fired with pure oxygen, leading to temperatures close to 3000 K. Since no soot is produced and no coal used, this flame is relatively easy to model (except for the common use of a 2D axisymmetric model, while the flame is contained in a square cross-section furnace). The modeling framework is similar to the one used by Pal et al. [27] and Mehta et al. [40], i.e., OpenFOAM together with a transported PDF. For chemistry the GRI-Mech 2.11 was employed, and the LBL-accurate PMC of Wang et al. [25] was used to determine radiation. They observed differences in computed mean temperatures of as much as 400 K when radiation is considered, and the results obtained with nongray radiation (including TRI) are in better agreement with experiment. They did not carry out optically thin-radiation or gray calculations; however, in this natural-gas-fired flame nongray effects should be strong. They did investigate the impact of TRI, and noted that, while there were considerable TRI in the flame itself, the overall impact of TRI was minimal. This is expected, since the majority of the furnace is filled with radiatively participating, nonreacting gases (see discussion in Sect. 6.1).

Edge et al. [68] and Clements et al. [69] studied the effects of radiation in coal burners running under oxy-fuel conditions. Both groups used both RANS and LES simulations employing the ANSYS Fluent CFD package, and both used Fluent's embedded (WSGG-based) gray spectral model as well as the FSCK spectral model, and both used Fluent's DOM RTE solver. Edge et al. [68] simulated firings from the RWE npower $0.5\,MW_{th}$ CTF at Didcot power station [70], using air (standard operation) or an O_2–CO_2 mixture (oxy-fuel operation). Apparently, the nongray FSCK model was only applied to converged RANS solutions, in order to predict radiative surface fluxes, and which showed considerable nongray effects (lowering fluxes by 20 % due to increased self-absorption). Clements et al. [69]

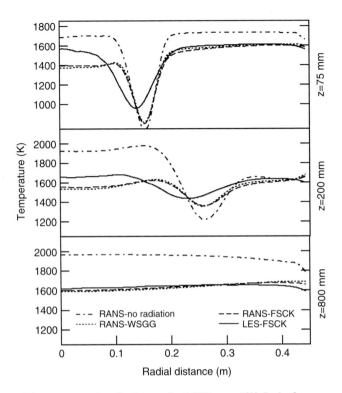

Fig. 6.12 Radial temperature plots for the oxy-fired CFD cases [69]. In the figure z represents the distance from the quarl exit

used a vertical down-fired cylindrical furnace, fitted with a scaled $250\,kW_{th}$ burner. They also considered both air and O_2–CO_2 mixture firings. Figure 6.12 shows resultingtemperature levels resulting from calculations without radiation, with gray radiation, and using the FSCK model. It is observed that in these strongly radiating coal flames radiation lowers temperature levels by up to $400\,K$, but there are virtually no differences between gray and nongray simulations. Only for wall-incident fluxes (not shown) was a slight nongray effect observed (lowering fluxes by less than 5 %). Further research is needed to see whether results from these select pilot scale facilities can also be applied to industrial plants.

6.6   Radiation Effects on Turbulence Levels

Little attention has been given to date to the effects thermal radiation may have on turbulence levels. Radiative transfer, through its "action at a distance," acts as a dissipative process, especially for large-scale turbulent structures with appreciable optical thickness. This was first noted by Townsend [71] for atmospheric turbulence, who combined analysis of the equations of mean fluctuations of velocity and temperature with the radiation field. He observed that appreciable radiative heat transfer

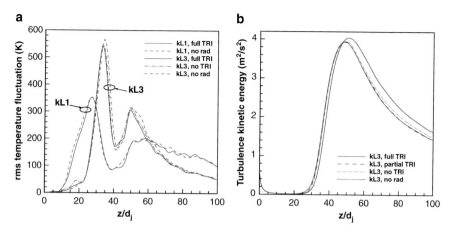

Fig. 6.13 Centerline profiles of root-mean-square (rms) temperature fluctuation and turbulence kinetic energy for Sandia D (kL1) and a scaled flame (kL3) [25]

leads to destruction of temperature fluctuations from the mean, to lower mean square temperature fluctuations, and also always to lower convection rates. The first to evaluate effects of radiation on turbulence in high-temperature combustion gases was perhaps Soufiani [72], who investigated the general case of homogeneous and isotropic turbulence in high-temperature H_2O and CO_2. The reduction in turbulence caused by radiation in an actual flame was first observed by Li [73], who calculated temperature fluctuations for his scaled flames with and without considering TRI, and found TRI to cause considerable decrease in $\langle T'^2 \rangle$-levels. Wang et al. [25] also calculated the effects of radiation feedback on the turbulent kinetic energy of their scaled flames (see Sect. 6.2.2). As shown in Fig. 6.13, in the larger flame the turbulent kinetic energy decreases by perhaps 5 % due to radiation, while TRI effects are almost negligible. Turbulence reduction of temperature fluctuations was generally small. Consalvi [54] also determined temperature fluctuations and turbulent kinetic energy for a methane pool fire (see Sect. 6.4), observing somewhat larger effects, indicating that radiation effects on turbulence may be more important in larger flames.

References

1. S. Mazumder, M.F. Modest, Turbulence–radiation interactions in nonreactive flow of combustion gases. J. Heat Transf. **121**, 726–729 (1999)
2. P.J. Foster, Relation of time-mean transmission of turbulent flames to optical depth. J. Inst. Fuel **42**(340), 179 (1969)
3. W. Krebs, R. Koch, H.J. Bauer, R. Kneer, S. Wittig, Effect of turbulence on radiative heat transfer inside a model combustor, in *Proceedings of Eurotherm Seminar No. 37—Heat Transfer in Radiating and Combusting Systems 2* (1994), pp. 349–362
4. S.J. Fischer, B. Hardoiun-Duparc, W.L. Grosshandler, The structure and radiation of an ethanol pool fire. Combust. Flame **70**, 291–306 (1987)

5. S.M. Jeng, M.C. Lai, G.M. Faeth, Nonluminous radiation in turbulent buoyant axisymmetric flames. Combust. Sci. Technol. **40**, 41–53 (1984)
6. J.P. Gore, G.M. Faeth, Structure and spectral radiation properties of turbulent ethylene/air diffusion flames, in *Proceedings of the Twenty-First Symposium (International) on Combustion* (1986), pp. 1521–1531
7. J.P. Gore, S.M. Jeng, G.M. Faeth, Spectral and total radiation properties of turbulent carbon monoxide/air diffusion flames. AIAA J. **25**(2), 339–345 (1987)
8. J.P. Gore, S.M. Jeng, G.M. Faeth, Spectral and total radiation properties of turbulent hydrogen/air diffusion flames. J. Heat Transf. **109**, 165–171 (1987)
9. J.P. Gore, G.M. Faeth, Structure and spectral radiation properties of luminous acetylene/air diffusion flames. J. Heat Transf. **110**, 173–181 (1988)
10. M.E. Kounalakis, J.P. Gore, G.M. Faeth, Mean and fluctuating radiation properties of non-premixed turbulent carbon monoxide/air flames. J. Heat Transf. **111**, 1021–1030 (1989)
11. Y. Zheng, R.S. Barlow, J.P. Gore, Measurements and calculations of spectral radiation intensities for turbulent non-premixed and partially premixed flames. J. Heat Transf. **125**, 678–686 (2003)
12. Y. Zheng, J.P. Gore, Measurements and inverse calculations of spectral radiation intensities of a turbulent ethylene/air jet flame, in *Thirtieth Symposium (International) on Combustion* (The Combustion Institute, Pittsburgh, PA, 2005), pp. 727–734
13. P.J. Coelho, Numerical simulation of the interaction between turbulence and radiation in reactive flows. Progr. Energy Combust. Sci. **33**, 311–383 (2007)
14. A.R. Masri, R.W. Bilger, R.W. Dibble, Turbulent nonpremixed flames of methane near extinction: mean structure from raman measurements. Combust. Flame **71**, 245–266 (1988)
15. A.R. Masri, R.W. Dibble, R.S. Barlow, The structure of turbulent nonpremixed flames revealed by Raman-Rayleigh-LIF measurements. Progr. Energy Combust. Sci. **22**, 307–362 (1996)
16. R.S. Barlow, J.H. Frank, Effects of turbulence on species mass fractions in methane/air jet flames. Proc. Combust. Inst. **27**, 1087–1095 (1998)
17. Combustion Research Facility, Sandia National Laboratories, in *International Workshop on Measurement and Computation of Turbulent Nonpremixed Flames.* http://www.sandia.gov/TNF/radiation.html
18. Y. Zheng, R.S. Barlow, J.P. Gore, Spectral radiation properties of partially premixed turbulent flames. J. Heat Transf. **125**, 1065–1073 (2003)
19. Y. Zheng, R.S. Barlow, J.P. Gore, Measurements and calculations of spectral radiation intensities for turbulent non-premixed and partially premixed flames. J. Heat Transf. **125**, 678–686 (2003)
20. G. Li, M.F. Modest, Application of composition PDF methods in the investigation of turbulence–radiation interactions. J. Quant. Spectrosc. Radiat. Transf. **73**, 461–472 (2002)
21. G. Li, M.F. Modest, Importance of turbulence–radiation interactions in turbulent diffusion jet flames. J. Heat Transf. **125**, 831–838 (2003)
22. G. Li, M.F. Modest, Numerical simulation of turbulence–radiation interactions in turbulent reacting flows, in *Modelling and Simulation of Turbulent Heat Transfer*, Chap. 3, ed. by B. Sundén, M. Faghri, (WIT, Southampton, 2005), pp. 77–112
23. P.J. Coelho, O.J. Teerling, D. Roekaerts, Spectral radiative effects and turbulence/radiation interaction in a non-luminous turbulent jet diffusion flame. Combust. Flame **133**, 75–91 (2003)
24. P.J. Coelho, Detailed numerical simulation of radiative transfer in a nonluminous turbulent jet diffusion flame. Combust. Flame **136**, 481–492 (2004)
25. A. Wang, M.F. Modest, D.C. Haworth, L. Wang, Monte Carlo simulation of radiative heat transfer and turbulence interactions in methane/air jet flames. J. Quant. Spectrosc. Radiat. Transf. **109**(2), 269–279 (2008)
26. A. Gupta, Large-eddy simulation of turbulent flames with radiation heat transfer. Ph.D. thesis, The Pennsylvania State University, University Park, 2011
27. G. Pal, A. Gupta, M.F. Modest, D.C. Haworth, Comparison of accuracy and computational expense of radiation models in simulation of nonpremixed turbulent jet flames. Combust. Flame **162**, 2487–2495 (2015)

28. Fluent, *FLUENT 6.0 UDF Manual* (Fluent Inc., New Hampshire, 2001)
29. *STAR-CD Computational Fluid Dynamics Software*, Version 6 (CD-adapco, New York, 2001)
30. B. Yang, S.B. Pope, An investigation of the accuracy of manifold methods and splitting schemes in the computational implementation of combustion chemistry. Combust. Flame **112**, 16–32 (1998)
31. S.R. Turns, *An Introduction to Combustion: Concepts and Applications*, 2nd edn. (McGraw-Hill, New York, 2000)
32. H. Jasak, A. Jemcov, Z. Tukovic, OpenFOAM: a C++ library for complex physics simulations, in *International Workshop on Coupled Methods in Numerical Dynamics* (IUC, Dubrovnik, 2007), pp. 1–20
33. A. Gupta, D.C. Haworth, M.F. Modest, Turbulence-radiation interactions in large-eddy simulations of luminous and nonluminous nonpremixed flames. Proc. Combust. Inst. **34**, 1281–1288 (2013)
34. B.R. Adams, P.J. Smith, Modeling effects of soot and turbulence–radiation coupling on radiative transfer in turbulent gaseous combustion. Combust. Sci. Technol. **109**, 121 (1995)
35. L. Tessé, F. Dupoirieux, J. Taine, Monte Carlo modeling of radiative transfer in a turbulent sooty flame. Int. J. Heat Mass Transf. **47**, 555–572 (2004)
36. A. Coppalle, D. Joyeux, Temperature and soot volume fraction in turbulent diffusion flames: measurements of mean and fluctuating values. Combust. Flame **96**, 275–285 (1994)
37. A. Soufiani, J. Taine, High temperature gas radiative property parameters of statistical narrow-band model for H_2O, CO_2 and CO, and correlated-k model for H_2O and CO_2. Int. J. Heat Mass Transf. **40**(4), 987–991 (1997)
38. G. Pal, M.F. Modest, A multi-scale full-spectrum k-distribution method for radiative transfer in nonhomogeneous gas–soot mixture with wall emission. Comput. Therm. Sci. **1**, 137–158 (2009)
39. R.S. Mehta, D.C. Haworth, M.F. Modest, Composition PDF/photon Monte Carlo modeling of moderately sooting turbulent jet flames. Combust. Flame **157**, 982–994 (2010)
40. R.S. Mehta, M.F. Modest, D.C. Haworth, Radiation characteristics and turbulence–radiation interactions in sooting turbulent jet flames. Combust. Theory Model **14**(1), 105–124 (2010)
41. R.S. Mehta, D.C. Haworth, M.F. Modest, An assessment of gas-phase thermochemistry and soot models for laminar atmospheric-pressure ethylene–air flames. Proc. Combust. Inst. **32**, 1327–1334 (2009)
42. C.K. Law, Comprehensive description of chemistry in combustion modeling. Combust. Sci. Technol. **177**, 845–870 (2005)
43. H. Chang, T.T. Charalampopoulos, Determination of the wavelength dependence of refractive indices of flame soot. Proc. R. Soc. (Lond.) A **430**(1880), 577–591 (1990)
44. J.H. Kent, D. Honnery, Modeling sooting turbulent jet flames using an extended flamelet technique. Combust. Sci. Technol. **54**, 383–397 (1987)
45. N.E. Endrud, Soot, Radiation and pollutant emissions in oxygen-enhanced turbulent jet flames. Master's thesis, The Pennsylvania State University, University Park, 2000
46. A. Wang, Investigation of turbulence–radiation interactions in turbulent flames using a hybrid FVM/particle-photon Monte Carlo approach. Ph.D. thesis, The Pennsylvania State University, University Park, 2007
47. S.-Y. Lee, Detailed studies of spatial soot formation processes in turbulent ethylene jet flames. Ph.D. thesis, The Pennsylvania State University, University Park, 1998
48. M. Klassen, Y.R. Sivathanu, J.P. Gore, Simultaneous emission absorption-measurements in toluene-fueled pool flames—mean and rms properties. Combust. Flame **90**, 34–44 (1992)
49. M. Klassen, J.P. Gore, Temperature and soot volume fraction statistics in toluene-fired pool fires. Combust. Flame **93**, 270–278 (1993)
50. A.Y. Snegirev, Statistical modeling of thermal radiation transfer in buoyant turbulent diffusion flames. Combust. Flame **136**, 51–71 (2004)
51. E. Gengembre, P. Cambray, D. Karmed, J. Belet, Turbulent diffusion flames with large buoyancy effects. Combust. Sci. Technol. **41**, 55–67 (1984)

52. J.M. Souil, P. Joulain, E. Gengembre, Experimental and theoretical study of thermal radiation from turbulent diffusion flames to vertical target surfaces. Combust. Sci. Technol. **41**, 69–81 (1984)
53. T.F. Smith, Z.F. Shen, J.N. Friedman, Evaluation of coefficients for the weighted sum of gray gases model. J. Heat Transf. **104**, 602–608 (1982)
54. J.L. Consalvi, Influence of turbulence–radiation interactions in laboratory-scale methane pool fires. Int. J. Therm. Sci. **60**, 122–130 (2012)
55. J.L. Consalvi, R. Demarco, A. Fuentes, Modelling thermal radiation in buoyant turbulent diffusion flames. Combust. Theory Model **16**(5), 817–841 (2013)
56. J.L. Consalvi, R. Demarco, A. Fuentes, S. Melis, J.P. Vantelon, On the modeling of radiative heat transfer in laboratory-scale pool fires. Fire Saf. J. **60**, 73–81 (2013)
57. S. Hostikka, K.B. McGrattan, A. Hamins, Numerical modeling of pool fires using LES and finite volume method for radiation, in *Proceedings of the Seventh International Symposium on Fire Safety Science*, ed. by D.D. Evans (International Association for Fire Safety Sciences, 2003), pp. 383–394
58. J. Cai, M. Handa, M.F. Modest, Eulerian–Eulerian multi-fluid methods for pulverized coal flames with nongray radiation. Combust. Flame **162**, 1550–1565 (2015)
59. J. Cai, X. Zhao, M.F. Modest, D.C. Haworth, Nongray radiation modelings in Eulerian–Lagrangian methods for pulverized coal flames, in *Paper No. TFESC-12950, Proceedings of the 1st Thermal and Fluids Engineering Summer Conference, TFESC-1*, New York (2015)
60. M. Syamlal, W. Rodgers, T. O'Brien, MFIX documentation: theory guide. Technical Note, DOE/METC-94/1004 (1993)
61. S.M. Hwang, R. Kurose, F. Akamatsu, H. Tsuji, H. Makino, M. Katsuki, Application of optical diagnostics techniques to a laboratory-scale turbulent pulverized coal flame. Energy Fuel **19**, 382–392 (2005)
62. M. Taniguchi, H. Okazaki, H. Kobayashi, S. Azuhata, H. Miyadera, H. Muto, T. Tsumura, Pyrolysis and ignition characteristics of pulverized coal particles. J. Energy Res. Technol. **123**(1), 32–38 (2001)
63. J. Cai, M.F. Modest, Absorption coefficient regression scheme for splitting radiative heat sources across phases in gas-particulate mixtures. Powder Technol. **265**, 76–82 (2014)
64. L. Chen, S.Z. Yong, A.F. Ghoniem, Oxy-fuel combustion of pulverized coal: characterization, fundamentals, stabilization and CFD modeling. Progr. Energy Combust. Sci. **38**, 156–214 (2012)
65. N. Nikolopoulos, A. Nikolopoulos, E. Karampinis, P. Grammelis, E. Kakaras, Numerical investigation of the oxy-fuel combustion in large scale boilers adopting the ECO-Scrub technology. Fuel **90**, 198–214 (2011)
66. M.F. Modest, *Radiative Heat Transfer*, 3rd edn. (Academic, New York, 2013)
67. X.Y. Zhao, D.C. Haworth, T. Ren, M.F. Modest, A transported probability density function/photon Monte Carlo method for high-temperature oxy–natural gas combustion with spectral gas and wall radiation. Combust. Theory Model **17**(2), 354–381 (2013)
68. P. Edge, S.R. Gubba, L. Ma, R. Porter, M. Pourkashanian, A. Williams, LES modelling of air and oxy-fuel pulverised coal combustion—impact on flame properties. Proc. Combust. Inst. **33**, 2709–2716 (2011)
69. A.G. Clements, S. Black, J. Szuhanszki, K. Stechly, A. Pranzitelli, W. Nimmo, M. Pourkashanian, LES and RANS of air and oxy-coal combustion in a pilot-scale facility: predictions of radiative heat transfer. Fuel **151**, 146–155 (2015)
70. L. Ma, M. Gharebaghi, R. Porter, M. Pourkashanian, J.M. Jones, A. Williams, Modelling methods for co-fired pulverised fuel furnaces. Fuel **88**, 2448–2454 (2008)
71. A.A. Townsend, The effects of radiative transfer on turbulent flow of a stratified fluid. J. Fluid Mech. **4**, 361–375 (1958)
72. A. Soufiani, Temperature turbulence spectrum for high-temperature radiating gases. J. Thermophys. Heat Transf. **5**(4), 489–494 (1991)
73. G. Li, Investigation of turbulence–radiation interactions by a hybrid FV/PDF Monte Carlo method. Ph.D. thesis, The Pennsylvania State University, University Park, 2002

Chapter 7
Radiative Heat Transfer in High-Pressure Combustion Systems

Many practical turbulent combustion systems operate at pressures higher than atmospheric. The general trend in propulsion applications, in particular (including piston engines [1], gas turbine combustors [2], and rockets [3]), is toward higher operating pressures. For example, peak pressures in modern heavy-duty diesel engines for trucks in the USA exceed 200 bar, and peak pressures in near term (~10 years, say) next-generation compression-ignition engines are expected to approach or exceed 300 bar. On the other hand, most detailed experimental data are limited to atmospheric pressure laboratory turbulent flames (Chap. 6), and relatively little work to date has focused on radiative heat transfer and TRI at elevated pressures. High-pressure specific issues in combustion and radiation heat transfer are reviewed briefly in Sect. 7.1. An example of a modeling study for a high-pressure laminar flame is provided in Sect. 7.2. Many practical combustion systems use fuels that are introduced as a liquid fuel spray, and spray/radiation coupling is discussed in Sect. 7.3. Recent work in modeling radiative heat transfer in piston engines is reviewed in Sect. 7.4, and examples for high-speed combustion systems are given in Sect. 7.5.

7.1 Combustion and Radiative Heat Transfer at Elevated Pressures

Chemical reaction rates $S_{\alpha,\text{chem}}$ for elementary reactions [Eqs. (2.3) and (2.4)] increase in proportion to the mixture pressure (reactant molar concentrations) raised to a power that is equal the number of molecules that must collide for the reaction to occur. Up to three-body reactions are of interest for combustion, so that reaction rates can increase in proportion to up to the third power of the pressure, although most gas-phase reactions of interest in combustion involve first-order (reaction rate

© The Author(s) 2016
M.F. Modest, D.C. Haworth, *Radiative Heat Transfer in Turbulent Combustion Systems*, SpringerBriefs in Applied Sciences and Technology,
DOI 10.1007/978-3-319-27291-7_7

proportional to p) or second-order (reaction rate proportional to p^2) reactions [4]. First-order and second-order reaction kinetics also are used in most models for soot formation and oxidation (Sect. 2.5.1). The total amount of soot that is present in a flame represents a balance between formation and oxidation rates. Across a wide range of combustion systems, it has been observed that the total amount of soot present (e.g., the peak soot volume fraction in the flame) scales approximately as p^n, where the value of n lies between 1.0 and 2.0 (depending on the specific system considered) for pressures ranging from 1 to $\sim$10 bar, and n falls to zero (soot "saturation") at higher pressures (above $\sim$40–50 bar).

Radiative emission (S_{emi}) and absorption (S_{abs}) rates for a participating molecular gas increase in proportion to the concentration of the participating species (hence in proportion to the mixture pressure p, for a given species mass or mole fraction), and the spectral properties also vary with pressure. In combustion applications spectral line broadening is governed by molecular collisions and is thus proportional to total pressure p. This leads to wider, more overlapping lines, making the gas "grayer." An example is given in Fig. 7.1, showing a small part of the spectrum for a mixture containing 1 % CO_2–2 % H_2O at varying total pressures. The areas under the curves are proportional to p, but it is seen that for higher pressure maxima decrease and minima increase due to line broadening. Still, even at 50 bar the absorption coefficient varies by a factor of 5 across this small part of the spectrum, leaving the gas strongly nongray, although line-by-line calculations may now require less than 100,000 spectral RTE solutions (as opposed to in excess of 1 million for 1 bar). Despite the increasing overlap, making the mixing models for k-distribution generation questionable, the FSK method using the uncorrelated mixture scheme, Eq. (3.72), was found to not lose accuracy (tested to 30 bar) [5]. It should also be noted here, that exact line broadening behavior at high pressures is not well understood, and extrapolation from atmospheric conditions is questionable.

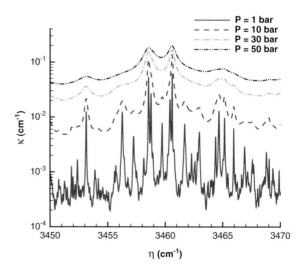

Fig. 7.1 Spectral absorption coefficient of a mixture containing 1 % CO_2–2 % H_2O at varying total pressures for a small region near 2.7 μm

Soot radiation depends on pressure only indirectly, through the soot volume fraction pressure dependence noted earlier, as well as secondary effects resulting (for example) from increased agglomeration of soot particles to form large chain-like aggregates at higher pressures.

7.2   A High-Pressure Laminar Flame

The influences of pressure on radiative heat transfer were explored for a two-dimensional (axisymmetric) steady laminar hydrogen–air diffusion flame in [6]. The configuration was similar to an inverse jet diffusion flame as described in Sect. 4.2.2 but with the central air jet and annular fuel jet issuing into an enclosure with a downstream converging chimney so that there was a large recirculation zone in the combustor. The simulations featured a nine species hydrogen–oxygen chemical mechanism and a mixture average molecular transport formulation. A full-spectrum k-distribution spectral model based on the HITEMP 2010 spectral database was used for H_2O radiation, and comparisons of results from different RTE solvers (spherical harmonics methods of different orders) were presented for three different mixture pressures: 1 bar, 5 bar, and 30 bar. Results from a line-by-line photon Monte Carlo method (based on the same underlying spectral database) and an optically thin model were also shown, for comparison purposes. At atmospheric pressure, the flame was found to be optically thin, and radiative emission resulted in only a small temperature drop. Radiation effects became more pronounced as the pressure increased, due to both increased emission/absorption and longer residence times (the reactant mass flow rates were the same for all three pressures). At higher pressure, the peak flame temperature was less affected by radiation because of the faster chemical reactions; however, due to larger residence time, significant cooling at downstream locations was observed. As the optical thickness increased with increasing pressure, differences between the spherical harmonics RTE solvers and the PMC solver became more evident. The general conclusion was that at elevated pressures, radiation effects are more prominent, and closer attention is required to the choice of RTE solver and spectral model.

7.3   Spray/Radiation Coupling

The extent to which spray/radiation effects (e.g., enhanced fuel vaporization rates due to absorption) are important in turbulent combustion is not known, and little work has been done in this area, especially for conditions that are of interest for practical combustion systems. One exception is an early numerical study of radiative transfer in a multiphase system that was intended to be representative of conditions in a diesel engine [7].

Here a recent study on coupling high-fidelity radiation models with fuel spray models in CFD is summarized [8], as an example of recent work in this area. The model was based on a conventional stochastic Lagrangian parcel method for the dispersed liquid phase (see Sect. 2.5.2). Key elements of the droplet radiation properties modeling included large, opaque droplets (so that geometric optics applied), cold droplets (neglecting droplet emission), gray droplets with a complex index of refraction corresponding to n-heptane in the mid-infrared, and droplet absorption efficiencies based on the normal incidence formulation. P_1 and PMC RTE solvers were implemented. Finite-volume cell-equivalent bulk radiative properties were deduced based on the local droplet distribution in each cell, and a consistent energy conserving scheme was used to distribute the radiative source term between the gas and liquid phases. The model first was validated for idealized statistically one-dimensional systems where analytic solutions were available, then was applied to an artificial spray flame (Fig. 7.2). The model was shown to accurately predict radiative sources even in cases of concentrated spray zones.

At the time of this writing, the models have been extended to consider more realistic optical and spectral properties of fuel droplets, and it has been found that the optical model, in particular, significantly affects the computed droplet radiation source term [S. Roy and W. Ge, personal communication]. It remains to be seen whether or not spray radiation plays a significant role in practical high-pressure combustion systems.

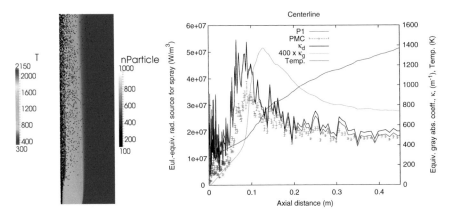

Fig. 7.2 *Left*: configuration of the artificial jet flame, with spray distribution overlaid on temperature contours. Only 4000 randomly chosen spray parcels are shown, for clarity. The quantity nParticle is the number of physical spray droplets in each parcel. *Right*: centerline profiles of total radiative heat source, droplet and gas absorption coefficients, and temperature [8]

7.4 Radiative Heat Transfer in Compression-Ignition Piston Engines

The importance of accurately accounting for realistic chemical kinetics and turbulence–chemistry interactions in engine simulations is well established (e.g., Fig. 1.1). On the other hand, while it has long been known that radiation can account for as much as 50 % of the total heat losses in very large bore, heavy-duty diesel engines [9], conventional wisdom had been that in-cylinder radiation was dominated by soot, and that radiation was of secondary importance in road vehicle scale engines (cars and trucks). Few in-cylinder CFD modeling studies have considered radiation heat transfer, and only a small subset of those has considered radiatively participating medium effects by solving an RTE. In [7], first- and third-order spherical harmonics methods were implemented with a delta-Eddington approximation to the scattering phase function for droplets, and solutions were presented for an axisymmetric finite-length cylinder with assumed spatial distributions of temperature, soot, and fuel droplets that were intended to be representative of those in a diesel engine during combustion. In [10], a discrete ordinates method (DOM) was implemented to explore the influence of radiation on NOx emissions in diesel engines; a significant reduction in computed NOx was found with consideration of soot radiation. A DOM method also was used in [11], with a wide band model for gas radiation and a gray model for soot. There spatial variations in wall temperature were considered. Quantitative comparisons were made with measured radiative heat fluxes, and the sensitivity of computed heat fluxes to the order of the DOM was studied; it was found that a low-order DOM was satisfactory in most cases. Similar radiation models were used in [12] to explore the effects of radiation on emissions. It was found that the influence of radiation on soot formation was larger than on NOx formation, and that wall temperature had a larger influence than radiation on emissions. In all of these CFD studies, the influences of turbulent fluctuations in composition and temperature on gas-phase chemistry, soot, and radiation (turbulence–chemistry–soot–radiation interactions) were neglected.

In the meantime, advanced high-efficiency engines that are currently being developed are expected to function close to the limits of stable operation [13], where even small perturbations to the energy balance can have a large influence on system behavior. The trends toward higher operating pressures and higher levels of exhaust gas recirculation (EGR) will make molecular gas radiation (absorption coefficient proportional to participating species concentration) more prominent. And increasing quantitative accuracy is being demanded from CFD-based models for in-cylinder processes, including accurate predictions of heat losses and pollutant emissions. For these reasons, radiative heat transfer under engine relevant conditions, including the effects of unresolved turbulent fluctuations, has been revisited recently. New analyses based on the most recent spectral property databases and high-fidelity radiative transfer equation (RTE) solvers show that at operating pressures and EGR levels typical of modern compression-ignition truck engines, radiative emission can be as high as 40 % of the wall heat losses, that molecular gas radiation (mainly CO_2

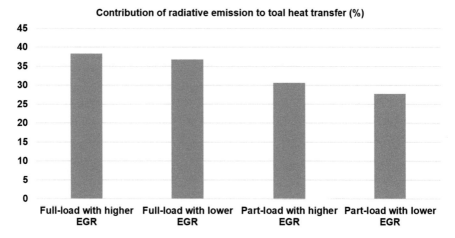

Fig. 7.3 Computed radiative emission fraction F_{rad} for a heavy-duty diesel engine at four operating conditions. Figure courtesy of V.R. Mohan

and H_2O) can be more important than soot radiation, that a large fraction of the emitted radiation (50 % or more) can be reabsorbed before reaching the walls, and that radiative heat transfer can be expected to result in significant changes in local in-cylinder temperatures (10 s of Kelvin) over engine relevant time scales. Examples for heavy-duty compression-ignition engines are discussed in the following.

In Fig. 7.3, the computed radiative emission from start of fuel injection through exhaust valve opening (total radiative energy emitted by the in-cylinder mixture through the combustion event, Q_{rad}), divided by the sum of the computed radiative emission plus boundary layer wall heat losses (total energy transfer from the in-cylinder gas to the wall through the combustion event computed using standard wall functions, Q_{wall}), is shown for a heavy-duty diesel engine at four operating conditions:

$$F_{rad} \equiv \frac{Q_{rad}}{Q_{rad} + Q_{wall}} . \tag{7.1}$$

The contribution from radiative emission ranges from approximately 25 % to approximately 40 %, depending on the operating conditions. The engine, operating conditions, and CFD modeling are described in [14]. The CFD results (obtained without radiation) were post-processed using an optically thin radiation model for CO_2, H_2O, and soot. Not all of the radiation that is emitted would reach the wall (as discussed further in the next example), but this suggests that radiation could contribute significantly to energy redistribution within the combustion chamber and/or to wall heat losses.

A second example for a heavy-duty optical diesel engine [15] at a part-load operating condition without EGR is provided in Fig. 7.4. This is an extension of the work reported in [16]. There an unsteady RANS formulation with a standard

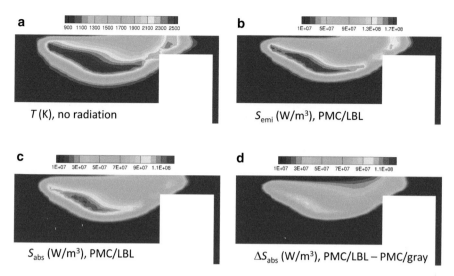

Fig. 7.4 Computed mean temperature, emission, and radiative absorption fields on a cutting plane corresponding to the injection axis for a heavy-duty optical diesel engine at an instant shortly after end of fuel injection. (**a**) Mean temperature without radiation. (**b**) Mean radiative emission rate for the PMC/LBL radiation model. (**c**) Mean radiative absorption rate for the PMC/LBL radiation model. (**d**) Difference in absorption rates between the PMC/LBL and PMC/gray radiation models. Figures courtesy of S. Roy and W. Ge

stochastic Lagrangian parcel spray model and a two-equation turbulence model with wall functions was used. A 34-species n-heptane chemical mechanism represented the chemistry, with no closure to account for effects of unresolved turbulent fluctuations in composition and temperature (a locally well-stirred reactor model at the finite-volume cell level), and with no soot model or radiation model. In [16], snapshots from the unsteady RANS simulations were post-processed to compute the local radiative source term (local rate of absorption minus rate of emission) using different spectral models and different RTE solvers, considering CO_2, H_2O, and CO radiation only. Radiation properties were based on the most recently available spectral radiation properties database (HITEMP2010), which covers pressures from 0.1 bar to 80 bar and temperatures from 300 to 3000 K. Simple pressure-based scaling was used to extrapolate to higher pressures, where needed. The results in Fig. 7.4 are from very recent fully coupled simulations, where the radiative source term feeds back into the CFD. A snapshot of the computed mean temperature field from a simulation without radiation is shown for reference, to give an indication of the degree of spatial nonhomogeneity in the temperature field. Local differences in computed mean temperatures at the instant shown for different combinations of RTE solvers and spectral models are as high as several tens of Kelvin (not shown), which is expected to influence computed emissions levels of NOx and soot. Computed radiative emission varies weakly for different combinations of RTE solver and spectral model, since changes in computed temperature and composition fields are relatively small. Radiative absorption (and hence the radiative source term) changes significantly, on the other hand, with both RTE solver and spectral model. The difference between LBL calculations and a gray model for the same PMC RTE solver is shown, as an example, in frame (d), indicating that the gray model vastly underpredicts self-absorption. The computed percentages of emitted radiation that are reabsorbed at the instant shown are 9.6 % for P1/gray, 60.8 % for P1/FSK, 25.9 % for PMC/gray, and 69.3 % for PMC/LBL. The rate of radiative heat loss to the wall (difference between emission and absorption) varies commensurately. Comparison between the (exact) PMC and P1 gray results indicates that P1 is not very accurate for the relatively optically thin conditions created by the gray assumption.

For this engine and operating condition, the total radiation heat loss to the wall over the combustion event is approximately 5 % of the boundary-layer convective wall heat loss. Radiation effects will be more prominent at higher loads (higher peak pressures) and for higher EGR rates.

These results do not consider soot radiation. Broadband soot radiation would significantly increase the local emission in locations of high soot volume fraction, and a smaller fraction of the emitted radiation would be reabsorbed by the participating gases (which absorb only in discrete wavenumber bands) before reaching the wall. This simplified analysis also neglects the influences of unresolved turbulent fluctuations in composition and temperature (turbulence–radiation interactions), which are expected to further enhance the overall level of radiative emission, as seen in the examples provided in earlier chapters. In general, then, it can be expected that radiation effects at this operating condition are significant, that reabsorption is important (requiring that the RTE be solved), that spectral radiation properties must be considered, and that radiation will significantly alter the in-cylinder temperature distribution in addition to contributing to heat losses.

The above observations may partially explain how CFD simulations that have ignored radiation altogether have been able to match experimental engine data. In the optically thick limit, a diffusion approximation would be an appropriate first-order model for molecular gas radiative heat transfer, and this effect could be mimicked by changing the apparent turbulent transport coefficient in the energy equation (e.g., the turbulent Prandtl number). As discussed in Sect. 5.2, it has been shown that radiation can significantly alter the turbulent boundary layer structure at high pressures [17, 18]. This may be an important consideration in engine applications, but again, to a first approximation this effect could be mimicked by modifying the parameters in the turbulence and/or wall heat transfer models. Truly predictive CFD-based models (that do not require tuning for each engine configuration and operating condition) will be needed to realize the ambitious efficiency and emissions targets that have been established for next generation engines and vehicles. To achieve that, it will be necessary to account correctly for each relevant physical process, rather than adjusting the model for one physical process (turbulent transport) to account for deficiencies in the modeling of a different physical process (radiative transport).

Significant work remains to be done to quantify radiation and TRI effects in engines over a wide range of operating conditions, and to develop appropriate models that can be used for predictive CFD simulations.

7.5 Radiative Heat Transfer in High-Speed Propulsion Systems

As mentioned earlier (Sect. 5.2), additional kinds of interactions related to compressibility can arise in high-speed systems. A number of studies have been published on radiative heat transfer in scramjet engines [19–22] and in rocket engines [23–27] and plumes [28], although none appear to have focused specifically on TRI or on compressibility-related interactions. One example of a scramjet modeling study and one example of a rocket engine modeling study focused on radiative heat transfer are discussed in the following.

In [20], a RANS-based modeling study of a combustion chamber similar to the HIFiRE-2 supersonic scramjet combustion chamber was presented. A 22-species chemical mechanism was used to model combustion of the methane–ethylene fuel/air oxidizer system. The simulations corresponded to a Mach 6.5 flight, where peak temperatures in the flame holder were approximately 2800 K and peak pressures at the flame holder exit were approximately 350 kPa. The combustion simulations were then post-processed to compute radiative intensity fields, using a simplified correlated-k narrow band spectral model, and two different methods to solve the RTE: DOM, and a ray-tracing method. Model results were compared with thermal radiative measurements from a corresponding experiment. Temperature drops due to radiation along various flow paths were estimated from the DOM results, and the predicted temperature reductions ranged from 2 to 245 K, with the largest temperature drops corresponding to fluid elements that followed streamlines into the flame holding cavity, where the residence times were the highest. The contribution of radiation (using the DOM model) to local wall heating was found to range from 1 to 10 % of the convective wall heating, depending on location (Fig. 7.5).

Radiative heat transfer was analyzed for subscale and full-scale rocket combustion chambers for H_2/O_2 and CH_4/O_2 reactants using the P_1 RTE solver together with various versions of weighted sum of gray gases models (WSGGMs) spectral models in [27]. An equilibrium chemistry model with a presumed mixture fraction PDF was used. The influences of variations in wall emissivities, different WSGGMs, the size of the combustion chamber, and of coupled (where radiation feeds back into the CFD) versus uncoupled (post-processed radiation) were investigated. The computed radiative wall heat flux varied by up to 30 % among the different spectral models for H_2/O_2 combustion, and by up to 50 % for CH_4/O_2 combustion (Fig. 7.6). The influence of radiation on the flow field was found to be negligible for the cases that were investigated, so that uncoupled radiation was sufficient. The local ratio of radiative to total wall heat flux was as high as 9–10 % for H_2/O_2 and as high as 8 % for CH_4/O_2.

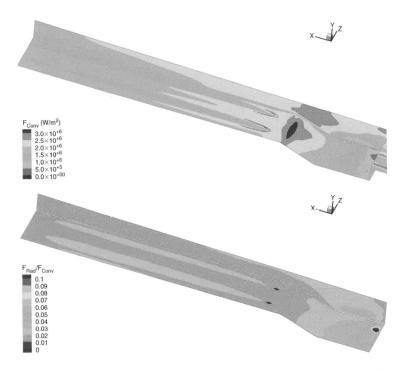

Fig. 7.5 Computed convective wall heating (**a**) and ratio of radiative to convective wall heating (**b**) from the uncoupled DOM model for a scramjet combustor [20]

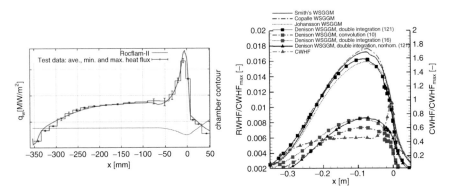

Fig. 7.6 Computed and measured wall heat flux profiles (*left*) and comparisons of computed normalized (by the maximum value of the computed convective wall heat flux, $CWHF_{max}$) profiles of radiative wall heat flux (RWHF) and convective wall heat flux (CWHF) for several different WSGGM spectral models (*right*) for the CH_4/O_2 subscale rocket combustion chamber [27]

As was the case for radiation in piston engines, much work remains to be done to understand radiation and TRI in high-speed combustion systems, and this is an area with great need for further research.

References

1. R.D. Reitz, Directions in internal combustion engine research. Combust. Flame **160**, 1–8 (2013)
2. A.H. Lefebvre, D.R. Ballal, *Gas Turbine Combustion: Alternative Fuels and Emissions*, 3rd edn. (CRC, Hoboken, 2010)
3. G.P. Sutton, O. Biblarz, *Rocket Propulsion Elements*, 8th edn. (Wiley, Hoboken, 2011)
4. S.R. Turns, *An Introduction to Combustion: Concepts and Applications*, 3rd edn. (McGraw-Hill, New York, 2011)
5. G. Pal, M.F. Modest, k-distribution methods for radiation calculations in high pressure combustion. J. Thermophys. Heat Transf. **27**(3), 584–587 (2013)
6. J. Cai, S. Lei, A. Dasgupta, M.F. Modest, D.C. Haworth, High fidelity radiative heat transfer models for high-pressure laminar hydrogen-air diffusion flames. Combust. Theor. Model. **18**, 607–626 (2014)
7. M.P. Mengüç, R. Viskanta, C.R. Ferguson, Multidimensional modeling of radiative heat transfer in diesel engines. SAE Technical Paper No. 850503 (1985)
8. S.P. Roy, J. Cai, M.F. Modest, Photon Monte Carlo method for radiation calculations in spray combustion, in *ICHMT International Symposium on Advances in Computational Heat Transfer*, Piscataway, 25–29 May 2015
9. G. Borman, K. Nishiwaki, Internal combustion engine heat transfer. Progr. Energy Combust. Sci. **13**, 1–46 (1987)
10. J. Abraham, V. Magi, Application of the discrete ordinates method to compute radiant heat loss in a diesel engine. Numer. Heat Transf. A **31**, 597–610 (1997)
11. J.F. Wiedenhoefer, R.D. Reitz, A multidimensional radiation model for diesel engine simulations with comparison to experiment. Numer. Heat Transf. A **44**, 665–682 (2003)
12. T. Yoshikawa, R.D. Reitz, Effect of radiation on diesel engine combustion and heat transfer. J. Thermal Sci. Technol. **4**, 86–97 (2009)
13. Office of Science and Office of Energy and Renewable Energy, U.S. Department of Energy, in *A Workshop to Identify Research Needs and Impacts in Predictive Simulation for Internal Combustion Engines (PreSICE)* (2011). http://www1.eere.energy.gov/vehiclesandfuels/pdfs/presice_rpt.pdf
14. V. Raj Mohan, D.C. Haworth, Turbulence-chemistry interactions in a heavy-duty compression-ignition engine. Proc. Combust. Inst. **35**, 3053–3060 (2015)
15. M.P.B. Musculus, P.C. Miles, L.M. Pickett, Conceptual models for partially premixed low-temperature diesel combustion. Progr. Energy Combust. Sci. **39**, 246–283 (2013)
16. S.P. Roy, J. Cai, A. Sircar, A. Imren, S. Ferreyro-Fernandez, D.C. Haworth, M.F. Modest, Radiative heat transfer under engine-relevant conditions, in *9th U.S. National Combustion Meeting*, Cincinnati, 17–20 May 2015
17. Y.F. Zhang, R. Vicquelin, O. Gicquel, J. Taine, Effects of radiation in turbulent channel flow: analysis of coupled direct numerical simulations. Int. J. Heat Mass Transf. **61**, 654–666 (2013)
18. R. Vicquelin, Y.F. Zhang, O. Gicquel, J. Taine, Effects of radiation in turbulent channel flow: analysis of coupled direct numerical simulations. J. Fluid Mech. **753**, 360–401 (2014)
19. H.F. Nelson, Radiative heating in scramjet combustors. J. Thermophys. Heat Transf. **11**, 59–64 (1997)
20. A.J. Crow, I.D. Boyd, M.S. Brown, J. Liu, Thermal radiative analysis of the HIFiRE-2 scramjet. AIAA Paper no. 2012-2751 (2012)
21. A.J. Crow, Computational uncertainty quantification of thermal radiation in supersonic combustion chambers. Ph.D. thesis, The University of Michigan, Ann Arbor, 2013
22. S.T. Surzhikov, J.S. Shang, Radiative heat exchange in a hydrogen-fueled scramjet combustion chamber. AIAA Paper no. 2013-0448 (2013)
23. B.E. Pearce, Radiative heat transfer within a solid-propellant rocket motor. J. Spacecraft Rockets **15**, 125–128 (1978)

24. K.J. Hammad, M.H. Naraghi, Radiative heat transfer in rocket thrust chambers and nozzles. AIAA Paper no. 89-1720 (1989)
25. R. Duval, A. Soufiani, J. Taine, Coupled radiation and turbulent multiphase flow in an aluminised solid propellant rocket engine. J. Quant. Spectrosc. Radiat. Transf. **84**, 513–526 (2004)
26. M.H. Naraghi, S. Dunn, D. Coats, Modeling of radiation heat transfer in liquid rocket engines. AIAA Paper no. 2005-3935 (2005)
27. F. Goebel, B. Kniesner, M. Frey, O. Knab, C. Mundt, Radiative heat transfer analysis in modern rocket combustion chambers. CEAS Space J. **6**, 79–98 (2014)
28. A.A. Alexeenko, N.E. Gimelshein, D.A. Levin, R.J. Collins, R. Rao, G.V. Candler, S.F. Gimelshein, J.S. Hong, T. Schilling, Modeling of flow and radiation in the Atlas plume. J. Thermophys. Heat Transf. **16**, 50–57 (2002)

Chapter 8
Summary, Conclusions, and Future Prospects

After many years of neglect or treatment with simplistic models, radiative heat transfer in general, and turbulence–radiation interactions in particular, are receiving increasing scrutiny in turbulent combustion research. It is recognized today that, in larger combustion systems of industrial and military relevance, radiation and its nonlinear interactions with turbulence and chemistry dominate the heat transfer in such flames. The rapidly growing body of literature that has been reviewed in this monograph is tangible evidence of this. Results from a wide range of experimental, modeling, and simulation studies show that radiation always reduces the temperature levels in reacting flows. This temperature drop is governed by the radiant fraction and may amount to hundreds of degrees Kelvin. Furthermore, turbulence–radiation interactions can be just as important as the turbulence–chemistry interactions that have been the subject of extensive research over several decades. While neglecting radiation in simulations leads to too high temperatures, using the optically thin approximation (OT) favored to date by many researchers (i.e., evaluating emission only, and neglecting absorption in the flame), will always lead to too low temperatures (provided emission TRI is included). Therefore, performing both types of calculations can bracket the importance of radiation in a given combustion system, and can thus be used to decide on the most appropriate radiation treatment.

Most research to date, be it experimental, analytical, or computational, has focused on small, open, atmospheric pressure jet and pool flames, both laminar and turbulent. Nonluminous laboratory-scale jet flames, such as Sandia D and others, tend to have small radiant fractions and, therefore, radiation effects are limited with maximum possible temperature drops of a few 10 s of degree Kelvin. In larger sooting flames discussed in this monograph, temperature drops larger than 300 K were observed, and generally about one third of that temperature drop was due to TRI.

© The Author(s) 2016
M.F. Modest, D.C. Haworth, *Radiative Heat Transfer in Turbulent Combustion Systems*, SpringerBriefs in Applied Sciences and Technology,
DOI 10.1007/978-3-319-27291-7_8

It was found that use of accurate nongray spectral models is more important than a sophisticated RTE solver, not only for nonluminous flames, but also for very sooty systems. Using gray spectral models always returns essentially OT results, even for very sooty flames, and can, therefore, not be recommended. Of the different spectral models available for conventional RTE solvers, the full-spectrum k-distribution method (FSK) probably offers the best compromise between accuracy and cost. For photon Monte Carlo (PMC) solutions, the line-by-line approach is found to be most suitable.

Three different RTE solvers show the most promise for reacting flow applications. The PMC method requires the largest CPU effort, but it is an exact method, it is the only one able to treat absorption TRI, and it can produce LBL spectral accuracy with little additional computational cost. When turbulence is treated stochastically, such as with a transported PDF method, it appears to be the natural choice, since it constitutes a transported PDF for photons and can be competitive in cost with conventional methods. The discrete ordinate method and its finite-volume cousin (DOM/FVM) is easy to program, but suffers from ray effects, and often high orders are required for reasonable accuracy. In addition, DOM/FVM requires iterations in scattering media and/or in the presence of reflecting walls, and convergence becomes difficult in optically thick media. The spherical harmonics method (SHM) is mathematically very involved and difficult to program. The lowest order P_1-approximation gives very good results in many applications at a very cheap price, but performs poorly in optically thin situations. Medium-order SHM generally outperforms DOM/FVM in optically medium-to-thick problems, especially in the presence of scattering and/or reflecting walls. Both conventional methods get more expensive when sophisticated spectral models, such as FSK, are employed. However, their cost may still be lower than PMC, in particular in transient problems.

Turbulence–radiation interactions were extensively investigated by several investigators. The simulations show that, if turbulence–radiation interactions are ignored, the radiative heat loss is always underpredicted and, consequently, temperature levels are generally overpredicted. Emission TRI were found to be always important, and need be considered even in small, optically thin flames (such as Sandia D), even if the OT model yields reasonable results. Absorption TRI, on the other hand, was found to be negligible in the vast majority of situations, making the optically thin fluctuation approximation (OTFA) an excellent tool. For very sooty flames, the absorption coefficient can be very large, and the adoption of this approximation is questionable. This is also true for molecular gases for a small part of the spectrum, since the spectral absorption coefficient at line centers can be very large. However, it was found that absorption TRI is negligible even in sooty systems if the flame is reasonably small (except for local effects), and that absorption TRI by strong lines never has a noticeable impact after integration over the entire spectrum. It was also found that, if LES simulations are used, absorption TRI may exist at the filter scale, but is always negligible at the subfilter scale, even for the optically thickest applications. Through freezing snapshots of turbulent flow fields the importance of different TRI-related terms was illuminated. In nongray media the absorption coefficient–Planck function correlation was found to be more important than other correlations and needs to be calculated accurately in order to determine turbulence–radiation interactions. Although nonlinearity of the Planck function on temperature is the severest among other functions, accurate determination of the temperature self-correlation alone appears insufficient in combustion gases (since emission takes place over narrow spectral intervals and is, therefore, not proportional to T^4).

Radiative heat transfer also has tremendous impact on the level of NOx production in a flame because of the strong temperature dependence of pollutant generation. It was found that use of different radiation model fidelity can lead to several orders-of-magnitude different NO level predictions. Comparison against experiment (Sandia D, Figs. 6.2 and 6.3) showed that, when choosing a high-fidelity radiation model, even rms values of NO can be predicted accurately. However, the reader is cautioned that radiation is only one of several difficult and important modeling tools (in addition to turbulence, chemistry, and soot models). Therefore, the choice of radiation model (i.e., accuracy vs. cost) should be compatible with the choices made for the remainder of the simulation.

As indicated earlier, most of the work to date has been limited to atmospheric pressure laboratory flames, while most practical combustion systems are larger, operate at elevated pressures and/or involve high-speed flow. As seen from the discussion in this monograph, very little work on the importance and effects of radiation and TRI have been carried out in piston engines, gas turbine combustors, rocket nozzles, etc., and much work remains to be done in these areas. Wall heat fluxes are of particular concern for combustor and injector durability in gas turbine combustors, scramjet engines, and rocket engines, and much of the radiation work that has been done to date has focused on the contribution of radiation to wall heat loads. In piston engines, the efficiency reduction associated with wall heat losses is a concern, and the radiative redistribution of energy within the combustion chamber (local temperature changes) can significantly influence pollutant emissions.

Finally, an area that has not been covered in this monograph, but that is expected to become increasingly important in the future, is the extension of radiative heat transfer models to interpret experimental data that are based on measurements of radiative intensities. With the increasingly sophisticated spectral models and RTE solvers that are becoming available, modelers can begin to directly compute the radiative intensity signals that correspond to various optical diagnostic techniques that are used in experiments. Thus the research community can move toward making direct comparisons of computed and measured radiative intensity signals that correspond to optical diagnostics techniques, rather than comparing quantities that are derived from those signals (e.g., species concentrations and temperature). This has the potential to significantly reduce uncertainty in comparing experimental and simulation results. At the same time, inverse methods are being developed that allow one to reconstruct the temperature field in a combustor (for example), based on an externally measured radiation signal. The combination of these two approaches has the potential to fundamentally alter the traditional process by which results from numerical simulations and from experimental measurements are compared, and is expected to ultimately provide tighter linkages among theory, numerical modeling, and experiment than has been the case to date.